Springer Theses

Recognizing Outstanding Ph.D. Research

Aims and Scope

The series "Springer Theses" brings together a selection of the very best Ph.D. theses from around the world and across the physical sciences. Nominated and endorsed by two recognized specialists, each published volume has been selected for its scientific excellence and the high impact of its contents for the pertinent field of research. For greater accessibility to non-specialists, the published versions include an extended introduction, as well as a foreword by the student's supervisor explaining the special relevance of the work for the field. As a whole, the series will provide a valuable resource both for newcomers to the research fields described, and for other scientists seeking detailed background information on special questions. Finally, it provides an accredited documentation of the valuable contributions made by today's younger generation of scientists.

Theses are accepted into the series by invited nomination only and must fulfill all of the following criteria

- They must be written in good English.
- The topic should fall within the confines of Chemistry, Physics, Earth Sciences, Engineering and related interdisciplinary fields such as Materials, Nanoscience, Chemical Engineering, Complex Systems and Biophysics.
- The work reported in the thesis must represent a significant scientific advance.
- If the thesis includes previously published material, permission to reproduce this must be gained from the respective copyright holder.
- They must have been examined and passed during the 12 months prior to nomination.
- Each thesis should include a foreword by the supervisor outlining the significance of its content.
- The theses should have a clearly defined structure including an introduction accessible to scientists not expert in that particular field.

More information about this series at http://www.springer.com/series/8790

Chao Zhang

Reliability of Steel Columns Protected by Intumescent Coatings Subjected to Natural Fires

Doctoral Thesis accepted by
Tongji University, Shanghai, China

 Springer

Author
Dr. Chao Zhang
National Institute of Standards
 and Technology
Gaithersburg, MD
USA

Supervisor
Prof. Guo-Qiang Li
Tongji University
Shanghai
China

ISSN 2190-5053   ISSN 2190-5061   (electronic)
Springer Theses
ISBN 978-3-662-52633-0   ISBN 978-3-662-46379-6   (eBook)
DOI 10.1007/978-3-662-46379-6

Printed on acid-free paper

Springer-Verlag GmbH Berlin Heidelberg is part of Springer Science+Business Media
(www.springer.com)

Declaration

The thesis and the research described and reported within has been completed solely by Chao Zhang under the supervision of Professor Guo-Qiang Li. Where other sources are quoted, full references are given.

May 2012   Chao Zhang

Publication

(Note: The publication list was updated on December 22, 2014)

International Journal Papers

1. C Zhang*, JL Gross, TP McAllister, GQ Li. Behavior of unrestrained and restrained bare steel columns subjected to localized fire. *Journal of Structural Engineering - ASCE*, 2014. DOI: 10.1061/(ASCE)ST.1943-541X.0001225. (*Corresponding author)
2. C Zhang, L Choe, M Seif, Z Zhang. Behavior of axially loaded steel short columns subjected to a localized fire. *Journal of Constructional Steel Research*, 2014. DOI:10.1016/j.jcsr.2014.11.012.
3. GB Lou, MC Zhu, M Li, C Zhang, GQ Li. Experimental research on slip-resistant bolted connects after fire. *Journal of Constructional Steel Research*, 2015;104:1–8.
4. C Zhang, GQ Li, YC Wang. Probabilistic analysis of steel columns protected by intumescent coatings to natural fires. *Structural Safety*, 2014;50:16–26.
5. C Zhang*, JL Gross, TP McAllister. Lateral torsional buckling of steel W-beams subjected to localized fires. *Journal of Constructional Steel Research*, 2013;88:330–338.
6. C Zhang*, GQ Li, A Usmani. Simulating the Behavior of Restrained Steel Beams to Flame Impingement from Localized-Fires. *Journal of Constructional Steel Research*, 2013;83:156–165.
7. GQ Li, C Zhang. The Chinese performance-based code for fire-resistance of steel structures. *International Journal of High-Rise Buildings*, 2013;2:123–130.
8. C Zhang, GQ Li, YC Wang. Predictability of buckling temperature of axially loaded steel columns in fire. *Journal of Constructional Steel Research*, 2012;75:32–37.

9. GQ Li, C Zhang*. Creep effect on buckling of axially restrained steel columns in real fires. *Journal of Constructional Steel Research*, 2012;71:182–188.

10. GQ Li, C Zhang*. Simple approach for calculating maximum temperature of insulated steel members in natural-fires. *Journal of Constructional Steel Research*, 2012;71:104–110.

11. GQ Li, C Zhang*, GB Lou, YC Wang, LL Wang. Assess the fire resistance of intumescent coatings by equivalent constant thermal resistance. *Fire Technology*, 2012;48:529–546.

12. C Zhang*, GQ Li, YC Wang. Sensitivity study on using different formulas for calculating the temperatures of insulated steel members in natural fires. *Fire Technology*, 2012;48:343–366.

13. C Zhang*, GQ Li. Thermal response of steel columns exposed to localized fires - numerical simulation and comparison with experimental results. *Journal of Structural Fire Engineering*, 2011;2:311–317.

14. C Zhang*, GQ Li. Fire dynamic simulation on thermal actions in localized fires in large enclosure. *Advanced Steel Construction*, 2012;8:124–136.

15. GQ Li, C Zhang*. Thermal response to fire of uniformly insulated steel members: background and verification of the formulation recommended by Chinese code CECS200. *Advanced Steel Construction*, 2010;6:788–802.

16. C Zhang*, GQ Li. Modified one-zone model for structural fire safety design and calculation. *Advanced Steel Construction*, 2013;9:282–297.

17. C Zhang, GQ Li, RL Wang. Using adiabatic surface temperature for thermal calculation of steel members exposed to localized fires. *International Journal of Steel Structures*, 2013;13:547–556.

18. C Zhang*, A Usmani. Thermal calculation of structures in fire: new insights from heat transfer principles. *Fire Safety Journal*, 2014. (Under Review)

19. C Zhang, JG Silva, C Weinschenk, D Kamikawa, Y Hasemi. Simulation methodology for coupled fire-structure analysis: modeling localized fire tests on a steel column. *Fire Technology*, 2014. (Submitted)

20. LL Wang, YL Dong, C Zhang. Experimental study on thermal insulating property of intumescent coatings exposed to large space fires. *Fire Technology*, 2014. (Submitted)

21. C Zhang. Simulating the behavior of a square hollow steel column subjected to localized fires. *Thin-Walled Structures*, 2014. (Submitted).

International Conference Papers

1. C Zhang*, JL Gross, L Choe. Behavior of steel components subjected to localized fires. *8th International Conference on Structures in Fire (SiF14)*, Shanghai, China, 2014, pp. 171–178. (Poster)

2. C Zhang, GQ Li, YC Wang. Effect of Aging on Reliability of Steel Columns Protected by Intumescent Coatings to Natural Fires. *7th International*

Conference on Structures in Fire (SiF'12), ETH Zurich, Switzerland, pp. 145–154. (Oral presentation)

3. C Zhang*, SC Jiang, GQ Li. Performance based fire resistance design of CFT columns in a railway station. *6th International Conference on Structures in Fire (SiF'10)*, MI, USA, 2010, pp. 441–448. (Poster)

4. GQ Li, C Zhang*. Thermal response of steel columns exposed to localized fires - numerical simulation and comparison with experimental results. *SiF10*, MI, USA, 2010, pp. 35–42. (Oral presentation)

5. GQ Li, C Zhang. Special Issues for Fire-Resistance of Steel Buildings. *HKIE 4th Annual Symposium 2012*, Hong Kong. (Invited keynote)

6. GQ Li, C Zhang. Fire resistance design on steel structures for large enclosure in China. *3rd International Symposium on Cold-formed Metal Structures*, Hong Kong, 2010, pp. 93–109. (Invited keynote)

7. C Zhang*, GQ Li, YZ Yin, MC Luo. Fire resistance design of large space grid structures by performance-based approach - a case study of the fire resistance design of the roof structure of Kunming International Airport. *6th International Conference on Advances in Steel Structures (ICASS'09)*, Hong Kong, 2009, pp. 776–785. (Oral presentation)

8. GQ Li, C Zhang. Fire resistance of restrained steel components. *4th International Conference on Protection of Structures against Hazards (PSH09)*, Beijing, China, 2009, pp. 21–38. (Invited keynote)

9. C Zhang*, GQ Li. Thermal behavior of a steel beam exposed to a localized fire - Numerical simulation and comparison with experimental results. *PSH09*, pp. 409–415. (Oral presentation)

10. GQ Li, C Zhang. Performance-Based Design of Restrained Steel Components against Fire. *Proceedings of State-of-the-Art Performance Based Design Method-Global Perspective and Overview*, Hong Kong, 2009, pp. 9–30. (Invited keynote)

Awards

- **IAFSS Best Thesis Award "Excellence in Research"** (years 2011–2014) by the International Association for Fire Safety Science (IAFSS).
  The award recognizes outstanding contributions to fire safety science and engineering. Three awards are made every 3 years worldwide.
- **Arup Research Price** by Ove Arup & Partners Hong Kong Ltd.
  The Prize awards postgraduate students to promote creativity and high standards of technology. I was the only recipient for the top prize for Mainland China in 2011.
- **Excellent Doctor Degree Dissertation of Tongji University 2013**.
- Fluor Scholarship 2010.
- Outstanding Graduate Student of Tongji University 2011.
- Outstanding Student of Tongji University 2010.
- Best paper award from China Association for Engineering Construction Standardization, 2010.

Supervisor's Foreword

The Ph.D. thesis hereinafter has comprehensively and deeply studied a new and very important problem in using intumescent coatings for protecting steel structures. In his cutting edge research, Dr. Zhang originally developed several theoretical models and engineering approaches for structural fire analysis and design. This Ph.D. thesis got a very high mark from the thesis defense committee and was nominated by the committee to receive the award of Excellent Doctor Degree Dissertation of the University in 2013. Because of his excellent work in this thesis, in 2014 Dr. Zhang received the prestigious IAFSS Best Thesis Award "Excellence in Research" by the International Association for Fire Safety Science. The award recognizes outstanding contributions to fire safety science and engineering. Three awards are made once every three years internationally.

Tongji University, December 2014   Guo-Qiang Li

Preface

In the past two decades, researchers from different countries have conducted series of experimental and theoretical studies to investigate the behavior of structures in fire. Many new insights, data, and calculation methods have been reported, which form the basis for modern interdisciplinary structural fire engineering. Some of these methods are now adopted in quantitative performance-based codes and have been migrated into practice. Difference between the observed behavior of structures in accident fires and the expected behavior determined by calculation methods based on the standard fire has increased interest in the performance of structures in natural fires but simultaneously questions the adequacy of both traditional and modern methodologies.

Intumescent coatings (IC) are passive fire protection materials widely used in steel construction. ICs are reactive materials, the behavior of which under heating condition is complex. The thermal insulation property of ICs cannot be measured by the current standard test methods which were originally developed for the traditional inert fireproofing materials. Besides, due to the organic components, ICs have aging problem. The reliability of structures protected by ICs in their service life is widely concerned by code authorities, fire bridges, engineers and coating manufacturers.

In a performance-based fire safety design through a rational approach, the risk of potential fire hazards on structures, which generally includes investigation of the fire occurrence probability, the failure probability (or reliability) of structures in fire, and the consequence of structural failure, should be assessed quantitatively. Steel columns are the most critical elements in a building, the failure of which usually leads to progressive collapse of the local or global structures. This research studies the reliability of steel columns protected by ICs in the potential fires occurred during their service life, and develops a probabilistic approach to determine the service life of ICs. A comprehensive investigation of all three aspects in structural fire engineering which include fire modeling, heat transfer analysis, and structural analysis is conducted, and several new theoretical models and engineering approaches are developed.

Post-flashover fires are considered as hazardous fires for structures. A modified one-zone model which considers the heat sink effect of steel members in the fire compartment by adding a quantity to the heat balance equation for the traditional one-zone compartment fire model is proposed, and an FE model is developed to simulate the modified one-zone model. The modified one-zone model can output both the fire (gas) temperature and the temperature of the steel members through a single calculation, and can yield more economical fire resistance design than the traditional one-zone model.

Background and shortcomings of current formulae for predicting steel temperature of insulated steel members are investigated. The current formulae are originally developed for calculation in standard fires so that they might give unacceptable results in natural fires, and the formulae are developed as simplified methods so that their applications are limited to situations where the properties of the insulation materials are or can be treated as constant or temperature-independent. Besides, when using the current formulae, iterative computations should be always processed, which is not convenient for the daily design work, and which is not efficient for probabilistic analyses which usually include hundreds of thousands of simulation loops.

A simple approach is proposed to calculate the maximum steel temperature of insulated steel members subjected to natural fires. The approach adopts time equivalent to relate natural fires with the standard fire, and uses a simple closed-form equation for calculating the maximum steel temperatures. The equation is obtained from curve fitting of the numerical data predicted by the program OZone V2, and is verified by the aforementioned modified one-zone model. The equation is also validated by test data. The equation gives satisfactory prediction of maximum steel temperatures in the range from 300 to 600 °C. The profession factor of the simple approach is characterized by test data, which has a mean of 0.955 and a COV of 0.014, and can be best described by a lognormal distribution. Professional factor is used to account for model error in calculation approaches, which is defined as the ratio of measured and predicted results.

ICs react at high temperatures and the thermal properties of ICs cannot be directly measured using the current standard test methods which are originally developed for traditional inert fireproofing materials. A simple procedure is proposed to assess the fire resistance of ICs by using the concept of equivalent constant thermal resistance. The procedure is based on an approximate formula for predicting the limiting temperature of protected steel members subjected to the standard fire. Test data from investigations on both small-scale samples and full-scale steel members are used to calculate the equivalent constant thermal resistance. Using the equivalent constant thermal resistance of ICs, the calculated steel temperatures agree well with the test data in the range of the limiting temperatures from 400 to 600 °C. Recent study on the effect of aging on the thermal insulation property of ICs is reviewed and the values of equivalent constant thermal resistance of ICs with different aging years are calculated.

The fundamental behavior of steel columns under fire conditions and experimental studies on steel columns in fire are reviewed. The accuracy and limitations of current calculation approaches to predict the buckling or critical temperature of steel columns are investigated by comparing with test data reported in the literature. The calculation approaches adopted in the Eurocode for predicting buckling temperature of steel columns are found to give acceptable prediction for tests with moderate utilization factor or load ratio, and give unacceptable prediction for tests with either high utilization factor ($\mu_0 > 0.83$) or low utilization factor ($\mu_0 < 0.16$). The professional factor for the simple closed-form equation in the Eurocode has a mean of 0.949 and a COV of 0.016, and can be best described by an extreme value distribution.

Monte Carlo simulations are conducted to get the reliability index or failure probability of steel columns protected by ICs in natural fires. The proposed simple closed-form equation is used to calculate the maximum steel temperature of the protected columns in natural fires and the simple closed-form equation in the Eurocode is used to calculate the column failure or buckling temperatures. The probability distributions of the investigated random parameters in the reliability analysis are determined from the literature. The effect of aging of ICs on the reliability index of protected steel columns is investigated. Case studies are conducted. The study finds that the aging of ICs has an effect of decreasing the reliability index of the protected steel columns in fire. That decreasing effect increases with increasing the aging years. The decreasing effect is more serious for columns with high load ratio than for columns with low load ratio. For the investigated cases with low load ratio ($\mu_0 \leq 0.3$), the amount of the decrease in reliability index due to aging is less than 0.2 and the corresponding increase in failure probability is less than 3 %. For the investigated cases with high load ratio ($\mu_0 > 0.3$), the maximum decrease in reliability index is about 0.24 and the corresponding maximum increase in failure probability is about 9 %. However, the ratio of failure probability change is large (can be about 45 %).

Based on the reliability analysis, a probabilistic approach is given to determine the service life of ICs for steel columns. In this approach, the failure probability of the protected steel columns is compared with the target probability of the structural fire design. The probability of fire occurrence and the probability of flashover are considered and are determined from codes. An example is given to determine the service life of ICs for protecting steel columns in an office building.

The approaches given in this study are also applicable for probabilistic analysis of steel columns protected by conventional inert fireproofing materials. Limited by test data on thermal insulation properties of intumescent coating with aging effects, the current study only considers steel columns protected by two different thicknesses of coatings. Also, many assumptions are made on the coating properties. Further studies using different coatings with various thicknesses are needed to obtain a general conclusion on the aging effect on reliability of intumescent coating protected steel columns in fire conditions. Suggestions for possible further work are made.

Acknowledgments

First of all I would like to thank my supervisor, Prof. Guo-Qiang Li for his incredible guidance, expert advice, kindness, and patience. I would also like to thank my co-supervisor Prof. Yong-Chang Wang at the University of Manchester for his support and valuable insight.

I would like to thank my industrial supervisor, Dr. Shou-Chao Jiang for his support and kindness. Thank you to Ms. Ya-Mei He and Dr. Guo-Biao Lou for their help for which I am very grateful. Thanks also to everybody at Research Group for Multi-Story and Tall Steel Buildings and Fire-Resistance of Steel Structures at Tongji University.

Thank you to my family, friends, and most of all Max for their faith and support throughout the last 4 years.

Contents

Chapter 1
Introduction

1.1 Background to the Project

In prescriptive codes, steel structures are commonly requested to be protected with thermal insulation to achieve the specified fire resistance ratings. The fire resistance rating of a building component is usually determined by the standard fire tests conducted on simply isolated members subjected to the standard fire exposures such as ISO834 [1]. The standard fire tests have been generally recognized to have many shortcomings, the two main aspects of which are on one hand the standard fire bears little resemblance to a real fire and on the other hand the behavior of isolated members can not represent the global behavior of structures in fire condition. Thus, the traditional prescriptive rules in codes, which, despite their relatively easy implementation, are inflexible and usually lead to expensive designs. As an alternative, performance-based codes have been developed in many countries for more reasonable fire safety design. In a performance-based fire safety design by a rational approach, the risk of real fire hazards on structures which generally includes a investigation of the fire occurrence probability, the failure probability (or reliability) of structures in fire, and the consequence of structural failure, should be assessed quantitatively [2, 3].

Intumescent coatings (IC), as passive fire protection materials, are widely used in industrial and public steel buildings [4] because of their advantages such as attractive appearance, potential for off-site application, and practically taking no space. The coatings, which are mostly composed of inorganic components contained in a polymer matrix, are inert at low temperatures and will expand and degrade to provide a charred layer of low conductivity materials at temperatures of approximately 280–350 °C [5, 6]. The charred layer, which acts as a thermal barrier, will prevent heat transfer to the underlying substrate. In practice, when specifying coating fire protection for steel structures, the designer assumes that the coating is correctly applied and its performance meets the fire protection needs without degradation over time. However, because of the organic components of intumescent coatings, the fire protection function of intumescent coatings over time would not be as reliable as when freshly applied [7].

© Springer-Verlag Berlin Heidelberg 2015
C. Zhang, *Reliability of Steel Columns Protected by Intumescent Coatings Subjected to Natural Fires*, Springer Theses, DOI 10.1007/978-3-662-46379-6_1

To date, some studies focused on reliability of structures in fire have been reported in literature. In 1980, Magnusson and Pettersson [8] reported the state-of-art of reliability studies in the area of fire-exposed members at the times. Magnusson [9] constructed a framework for probabilistic analysis of fire exposed steel structures. In his study, the real fires are represented by the "Swedish" fire curves derived from the widely used one zone compartment fire model. Woeste and Schaffer [10] reported a reliability analysis of fire exposed wood joist floor assemblies by using second moment approximations. In their study, the fire severity was considered using a standard fire duration time calculated from a "t-equivalent" formula developed for ventilation controlled fires. The concept of "t-equivalent" was originally proposed by Ingberg [11] to relate real fires with the standard fire. So far, several formulae, which are based on different assumptions, have been developed for "t-equivalent" calculations [12, 13]. The concept is adopted by the probabilistic model for fire load given by JCSS (Joint Committee on Structural Safety) [14]. Recently, He and Grubits [15] used the concept to assess the failure probability of building structures. Lange et al. [16] reported a reliability analysis of a composite floor slab in fire by using Monte Carlo Simulations (MSC). The fire model used was the Eurocode parametric fire model [17]. Hietaniemi [18] reported a probabilistic simulation of fire endurance of a wooden beam. In his study, the sophisticated CFD model, FDS (Fire Dynamics Simulator [19]), was adopted for fire modeling.

However, study on reliability of steel structures protected by ICs in fire conditions has not been reported in literature. Consider the wide usage of ICs, the reliability of structures protected by ICs in their service life is concerned by code authorities, fire bridges, engineers and coating manufacturers.

1.2 Objective and Originality

The objective of this project is to study the reliability of steel columns protected by ICs subjected to natural fires and develop a probabilistic approach to determine the service life of ICs. To achieve this objective, A comprehensive investigation on all three aspects in structural fire engineering which include fire modeling, heat transfer analysis and structural analysis is conducted, and several new theoretical models and engineering approaches are developed:

- A modified one zone model is proposed for structural fire analysis. A FEM heat transfer model using ANSYS is developed to solve the heat balance equation of the modified one zone model.
- A simple approach is developed to calculate the maximum temperature of insulated steel members subjected to natural fires. The simple model is verified by program OZone V2 and the proposed modified one zone model. The simple model is also validated by test data, and the model error or profession factor of the simple model is characterized using the test data.

- A simple procedure is proposed to assess the fire resistance of ICs. The procedure is based on the concept of equivalent constant thermal resistance. The constant thermal resistance of ICs with different aging years is derived.
- The accuracy and limitations of calculation approaches for predicting the buckling or critical temperature of steel columns are investigated by comparing with test data reported in literature. The professor factors of the approaches are characterized by test data.
- The effect of aging of ICs on the reliability of protected steel columns in fire is investigated. Monte Carlo simulation method is used to calculate the failure probability and reliability index.
- A probabilistic approach is developed to determine the service life of ICs for steel columns. A example is given to use the approach ro determine the service life of ICs for protecting steel columns in an office building.

1.3 Outline of Chapters

The thesis includes 8 chapters plus 4 appendixes, which are:

Chapter 1 Introduction

Chapter 2 Fire Modeling The widely used one zone model for modeling post-flashover fires is presented. Detailed description of the sub-models in the one zone model is given. A FE model is developed in ANSYS to simulate the one zone model to give temperature-time curves for post-flashover fires.

Chapter 3 Steel Temperature in Natural Fires The one-dimensional heat transfer model used for calculating the temperature of steel members insulated by coatings is introduced. Background for current formulae given by fire codes in different countries are introduced and shortcomings of current formulae are presented. A modified one zone model has been proposed to calculate the steel temperature in natural fires. A FE model has been developed to solve the heat balance equation of the modified one zone model. A simple closed-form expression, which only needs hand calculation, has been proposed for calculating the maximum steel temperature of insulated steel members in natural fires. The probabilistic distribution of the profession factor for the proposed simple expression has been characterized by test data.

Chapter 4 Thermal Properties of Intumescent Coatings in Fire The behavior of intumescent coatings under heating is studied. Mathematical models to study the heat transfer of ICs, and current approaches to assess the thermal insulation properties of ICs have been reviewed. A simple procedure has been developed to assess the fire resistance of intumescent coatings by using the concept of equivalent constant thermal resistance. The concept and derivation of the constant thermal resistance has been presented. The simple procedure has been validated by test data on both small and full scale samples. The constant thermal resistance of intumescent coatings with aging effect has been calculated.

Chapter 5 Behavior of Steel Columns in Fire The fundamental behavior of steel columns in fire conditions isgiven. Different calculation approaches for predicting the buckling or critical temperature of steel columns are presented. The accuracy and limitations of those approaches are determined by comparing with test data reported in literature. The professional factors for those approaches are also characterized and the probabilistic distributions of the professional factors are derived.

Chapter 6 Reliability Analysis Reliability analysis of steel columns protected by ICs subjected to natural fires is conducted. Statistics of random parameters are determined from literature. Monte Carlo simulation method is used to calculate failure probability and reliability index. Different cases are considered. The values of failure probability and reliability index for steel columns protected by intumescent coating with different aging years are calculated.

Chapter 7 Service Life of Intumescent Coatings A probabilistic approach is given to determine the service life of ICs for steel columns. A example is given to show the application of the approach.

Chapter 8 Conclusions and Further Work

Consider the importance of heat transfer in structural fire engineering, and for easy understanding of the thesis to readers who lack a knowledge of heat transfer, Appendix A, is included to introduce the fundamental principles of heat transfer in structural fire analysis. Appendix B gives different material models for structural steel at elevated temperatures. Appendixes C and D are commands used in the thesis.

References

1. B. 476-20, *Fire Tests on Building Materials and Structures, Part 20: Methods for Determination of the Fire Resistance of Elements of Construction (General Principles)* (British Standards Institution, London, 1987)
2. CIBW014, *Rational Fire Safety Engineering Approach to Fire Resistance of Buildings*. Technical report (2001)
3. P. 7974-7, *Application of fire safety engineering principles to the design of buildings–part 7: Probabilistic risk assessment*. Technical report (2003)
4. Corus, *Fire Resistance of Steel-Framed Buildings*, 2006 edition. (Corus Construction and Industrial, 2006)
5. S. Bourbigot, S. Duquesne, J. Leroy, J. Fire Sci. **17**, 42 (1999)
6. M. Gillet, L. Autrique, L. Perez, J. Phys. D: Appl. Phys. **40**, 883 (2007)
7. L. Wang, Y. Wang, G. Li, in *Proceedings of the Sixth International Conference on Structures in Fire* (MI, 2010), pp. 735–742
8. S. Magnusson, O. Pettersson, Fire Saf. J. **3**, 227 (1980)
9. S. Magnusson, *Probabilistic analysis of fire exposed steel structures*. Technical report (1974)
10. F. Woeste, E. Schaffer, Fire Mater. **3**, 126 (1979)
11. S. Ingberg, NFPA Q. **22**, 43 (1928)
12. T. Harmathy, Fire Mater. **11**, 95 (1987)
13. V. Kodur, M. Dwaikat, Mater. Struct. **43**, 1327 (2010)
14. JCSS, *JCSS probabilistic model code, part ii–load models*. Technical report (2001)
15. Y. He, S. Grubits, J. Fire Protect. Eng. **20**, 5 (2010)
16. D. Lange, A. Usmani, J. Torero, in *Proceedings of the Fifth International Conference on Structures in Fire* (Singapore, 2008), pp. 760–770

17. BSI, *Eurocode 1: Actions on Structures–Part 1–2: General Rules–Actions on Structures Exposed to Fire* (British Standard, London, 2002)
18. J. Hietaniemi, Struct. Saf. **29**, 322 (2007)
19. K. McGrattan, R. McDermott, S. Hostikka, J. Floyd, *Fire dynamics simulator (version 5) user's guide*. Technical report, NIST Special Publication 1019-5. NIST (2010)

Chapter 2
Fire Modeling

2.1 Introduction

A real fire will generally undergo six stages which include ignition, growth, flashover, full fire development or steady burning, decay and extinguishment. Flashover is the rapid transition between the primary fire which is essentially localized around the item first ignited, and the general conflagration within the compartment when all fuel surfaces are burning [1]. Depending on whether flashover will happen or not, the real fires are usually divided into pre- and post-flashover fires. For small and middle scaled compartments with sufficient fuel and ventilation, the potential fires will develop to flashover and be characterized as post-flashover fires. For large scale enclosures or where sprinklers work effectively, flashover is unlikely to occur and the fires are characterized as pre-flashover fires. Post-flashover fires provide the worst case scenario which are usually considered in fire resistance design. However, localized heating of key elements of structure in pre-flashover fires must also be considered.

The behavior of a real fire is complex, which depends on many parameters such as active fire detection and suppression systems (smoke detector and sprinkler), fire load (amount and distribution), combustion, ventilation, compartment size and geometry, and thermal properties of compartment boundaries [2]. So far, with increase in complexity, empirical correlations (e.g. nominal fire curves and parameter fire curve [3]), zone models (e.g. one-zone models for post-flashover fires [4, 5] and two-zone models for pre-flashover fires [5, 6]), and sophisticated CFD models (e.g. Fire Dynamics Simulator [7]) have been developed to model the fire behavior. Also, stochastic models have been developed for compartment fires [8, 9].

In this chapter, the widely used one-zone model for modeling post-flashover fires is presented. Detailed description of the sub-models in one-zone model is given. A FE model is developed in ANSYS to simulate the one-zone model to give temperature-time curves for post-flashover fires. The concept of equivalent fire severity is also reviewed.

© Springer-Verlag Berlin Heidelberg 2015
C. Zhang, *Reliability of Steel Columns Protected by Intumescent Coatings Subjected to Natural Fires*, Springer Theses, DOI 10.1007/978-3-662-46379-6_2

2.2 One-Zone Compartment Fire Model

2.2.1 Heat Balance Equation

Figure 2.1 shows the widely used one-zone model for post-flashover fires. In this model, the heat balance equation within the compartment is given by [10]

$$HRR = \dot{q}_g + \dot{q}_w + \dot{q}_{o,c} + \dot{q}_{o,r} \tag{2.1}$$

where HRR is heat release rate due to combustion; $\dot{q}_g$ is rate of heat storage in the gas volume; $\dot{q}_w$ is rate of heat loss through the walls, ceiling and floor; $\dot{q}_{o,c}$ is rate of heat loss due to replacement of hot gases by cold; and $\dot{q}_{o,r}$ is rate of heat loss by radiation through the openings. Based on this model, both analytical expressions and compute programs have been developed to give fire curves for structural design [11].

2.2.2 Heat Release Rate (HRR)

Heat release rate (HRR) is the most important variable in measuring fire severity, which can be calculated by

$$HRR = \dot{m}_f \cdot \Delta H_c \tag{2.2}$$

where, $\dot{m}_f$ is the mass burning rate of the fuel; and ΔH_c is the net heat of combustion of the fuel. In ventilation controlled fires (fully-developed compartment or post-flashover fires), the HRRs can be alternatively calculated by [1]

$$HRR = \dot{m}_{air} \cdot \Delta H_{air} \tag{2.3}$$

where, $\dot{m}_{air}$ is the mass rate of air inflow and ΔH_{air} is the heat released per unit mass air consumed, for most common fuel, $\Delta H_{c,air} = 3.03 \pm 0.02$ MJ/kg [1].

Fig. 2.1 Illustration of compartment one-zone model

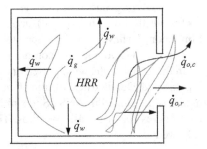

Fig. 2.2 Illustration of the HRR history in a NFSC fire

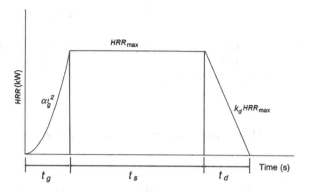

The HRR of a real fire can be measured by cone calorimeter [12]. In design work, the natural fire safety concept (NFSC) is widely used to represent the fire conditions [3, 5]. As shown in Fig. 2.2, the NFSC fire is assumed to be t-square in the growth stage and decay stage begins at the time when 70 % of design fire load is consumed. In a NFSC fire, at the growth stage, the HRR is given by

$$HRR = \alpha t^2 \tag{2.4}$$

the fire growth time t_g is given by

$$t_g = \sqrt{\frac{HRR_{max}}{\alpha}} \tag{2.5}$$

and the fuel energy consumed at the fire growth stage, Q_g, is

$$Q_g = \int_0^{t_d} \alpha t^2 dt = \frac{\alpha t_g^3}{3} \tag{2.6}$$

where, α is the fire intensity coefficient, taken as 0.00293, 0.0117 and 0.0466 for slow, medium and fast growth fires, respectively.

The duration time of steady burning in a NSFC fire is given by

$$t_s = \frac{0.7 q_f A_f - Q_g}{HRR_{max}} \tag{2.7}$$

and the duration of decaying stage is given by

$$t_d = \frac{0.6 q_f A_f}{HRR_{max}} \tag{2.8}$$

where, q_f, A_f are design fire load density and floor area, respectively; and HRR_{max} is the maximum heat release rate, for ventilation controlled fires [1],

$$HRR_{\max} = 1,500 A_o \sqrt{H_o} \tag{2.9}$$

where, A_o is the total area of openings; and H_o is the weighted average of opening heights. For fuel-controlled fires, EC1 [3] gives some values of maximum heat release rates for different occupancies.

2.2.3 Heat Loss Due to Convective Flow ($\dot{q}_{o,c}$)

Pressure in a compartment fire is essentially atmospheric, and flows occur at openings due to hydrostatic pressure differences. The mass rates of out and in flows are calculated by [1]

$$\dot{m}_{gas} = \frac{2}{3} C_d W_o \rho_\infty [2g \frac{T_\infty}{T_g}(1 - \frac{T_\infty}{T_g})]^{1/2}(H_o - X_N)^{3/2} \tag{2.10}$$

and

$$\dot{m}_{air} = \frac{2}{3} C_d W_o \rho_\infty [2g(1 - \frac{T_\infty}{T_g})]^{1/2}(X_N - X_d)^{1/2}(X_N + X_d/2) \tag{2.11}$$

respectively. The heat loss due to replacement of hot gases by cold at openings is calculated by

$$\dot{q}_{o,c} = \dot{m}_{gas} c_p T_f - \dot{m}_{air} c_p T_\infty \tag{2.12}$$

The flow coefficient, C_d, is approximately 0.7. The neutral plane height, X_N, can be approximated as the half height of the opening. For post-flashover fires, the height of the interface X_d is approximately zero. Taken those values into Eq. 2.11, the inflow mass rate for post-flashover fires can be approximated as

$$\dot{m}_{air} = 0.5 A_o \sqrt{H_o} \tag{2.13}$$

In many calculations [1], the mass rate of outflow is taken as equal to the mass rate of inflow given by Eq. 2.13. Correspondingly, the heat loss due to replacement of hot gases by cold is calculated by

$$\dot{q}_{o,c} \approx 0.5 A_o \sqrt{H_o} c_p (T_f - T_\infty) \tag{2.14}$$

where, c_p is the specific heat of the gas; T_f and T_∞ are fire and environment temperatures, respectively.

2.2.4 Heat Loss by Radiation Through the Openings ($\dot{q}_{o,r}$)

The heat loss by radiation through the openings of a fire compartment is usually calculated by adopting homogeneous gray gas approximations for fire environment and using Stefan-Boltzmann law, that,

$$\dot{q}_{o,r} = A_o \varepsilon_f \sigma (T_f^4 - T_\infty^4) \tag{2.15}$$

where, ε_f is the effective emissivity of the gases within the compartment, which can be calculated from

$$\varepsilon_f = 1 - e^{-\kappa L} \tag{2.16}$$

where, κ is the extinction coefficient, usually taken as $1.1\,\mathrm{m}^{-1}$ [10]; and L is the mean beam length for the enclosure, which is approximated as the height of the compartment.

2.2.5 Heat Loss Through Wall ($\dot{q}_w$)

The heat transfer into the boundary surface of a compartment occurs by convection and radiation from the enclosure, and then conduction through the walls. In calculating heat transfer from fire to boundary surface ($\dot{q}_w$), the following assumptions are adopted,

- In radiation calculation, the fire and the boundary surface are represented as two infinitely parallel grey planes that the view factor is taken as unit;
- In unsteady conduction calculation, The boundaries (walls, ceiling and floor) are usually assumed to be semi-infinite solids.

Theoretically, for semi-infinite behavior, the exposure time must be less than the penetration time (which for a slab exposed to Dirichlet boundary condition is about $\frac{1}{\alpha_w}(\frac{\delta_w}{4})^2$ [1]). In practice, if the thickness of a slab is greater than $2\sqrt{\alpha_w t}$, the semi-infinite solid assumption is applicable [1, 13].

The governing equation for 1D wall conduction is given by

$$\frac{\partial^2 T_w(x,t)}{\partial x^2} = \frac{1}{\alpha_w} \frac{\partial T_w(x,t)}{\partial t} \tag{2.17}$$

where, $\alpha_w = k_w/(\rho_w c_w)$ is the thermal diffusivity of the wall. k_w, ρ_w and c_w are conductivity, density and specific heat of the wall, respectively.

At fire and environment exposed sides, the Neumann boundary conditions are given by

$$\dot{q}_w = -k_w \frac{\partial T_w(0,t)}{\partial x} = (h_{c,l} + h_{r,l})[T_f - T_w(0,t)] \tag{2.18}$$

and

$$-k_w \frac{\partial T_w(\delta_w, t)}{\partial x} = (h_{c,r} + h_{r,r})[T_w(\delta_w, t) - T_\infty], \qquad (2.19)$$

respectively. Here, $T_w(0, t)$ and $T_w(\delta_w, t)$ are temperatures of fire and environment exposed surfaces, respectively; δ_w is the thickness of the wall; $h_{c,l}$ and $h_{c,r}$ are convective heat transfer coefficients at fire and environment exposed surfaces, taken as 35 and 9 W/m^2K [3], respectively; and $h_{r,l}$ and $h_{r,r}$ are radiative heat transfer coefficients at fire and environment exposed surfaces, given by

$$h_{r,l} = \varepsilon_{res}\sigma[(T_f + 273)^2 + (T_w(0, t) + 273)^2][T_f + 273 + T_w(0, t) + 273] \quad (2.20)$$

and

$$h_{r,r} = \varepsilon_w\sigma[(T_\infty + 273)^2 + (T_w(\delta_w, t) + 273)^2][T_\infty + 273 + T_w(\delta_w, t) + 273] \quad (2.21)$$

respectively. Here, ε_{res} is the resultant emissivity at fire exposed surface, given by

$$\varepsilon_{res} = \frac{1}{1/\varepsilon_f + 1/\varepsilon_w - 1} \qquad (2.22)$$

in which, ε_w is the emissivity of the wall.

2.2.6 Heat Storage in Gas Volume ($\dot{q}_g$)

The heat stored in the gas volume within the compartment is calculated by

$$\dot{q}_g = \rho_\infty c_p V_g \frac{\Delta T_f}{\Delta t} \qquad (2.23)$$

where, ρ_∞ is the density of the gas; and V_g is the gas volume.

2.3 Temperature-Time Curves for Post-Flashover Fires

2.3.1 Standard Fire

The ISO834 standard temperature-time curve is given by [3]

$$T_g = 20 + 345\log_{10}(8t + 1) \qquad (2.24)$$

where, t is the standard fire exposure time in minutes.

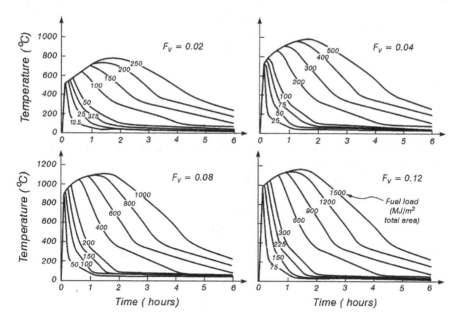

Fig. 2.3 Swedish *curves*

2.3.2 Swedish Curves

As shown in Fig. 2.3, the 'Swedish' fire curves are widely referenced in fire engineering. The 'Swedish' fire curves are derived by solving the heat balance equation for one-zone compartment fire model (see Eq. 2.1) with using the burning rate of ventilation controlled fires [10].

2.3.3 Lie Correlations

In SFPE handbook [14], correlations given by Lie are adopted for determining real fire temperatures. In the heating phase, the expression is given by

$$T_g = 250(10F_o)^{\frac{0.1}{F_o^{0.3}}} e^{-2F_o t}[3(1 - e^{-0.6t}) - (1 - e^{-3t})$$
$$+ 4(1 - e^{-12t})] + C(\frac{600}{F_o})^{0.5} \tag{2.25}$$

where, C is a constant taking into account the influence of the properties of the boundary material on the temperature. $C = 0$ for heavy materials ($\rho \geq 1,600\,\text{kg/m}^3$), and $C = 1$ for light materials ($\rho < 1,600\,\text{kg/m}^3$). F_o is the opening factor, defined by

$$F_o = \frac{A_o\sqrt{H_o}}{A_T} \tag{2.26}$$

Equation 2.25 is valid for

$$t \leq \frac{0.08}{F_o} + 1 \tag{2.27}$$

and

$$0.01 \leq F_o \leq 0.15. \tag{2.28}$$

If $t > 0.08/F_o + 1$, a value of $t = 0.08/F_o + 1$ should be used. If $F_o > 0.15$, a value of $F_o = 0.15$ should be used.

The duration of heating is determined by

$$\tau = \frac{q_{t,wood}}{330 F_o} \tag{2.29}$$

where, $q_{t,wood}$ is the wood equivalent fire load density, determined by

$$q_{t,wood} = \frac{q_f A_f}{A_t} \frac{1}{\triangle H_{c,wood}} \tag{2.30}$$

here, A_t is the area of bounding surfaces, taken as the total enclosure area in post-flashover fires; and $\triangle H_{c,wood}$ is the net calorific value of wood, often taken as 17.5 MJ/kg [3].

In the cooling phases, the fire temperature is calculated by

$$T_g = -600(\frac{t}{\tau} - 1) + T_\tau \geq 20\,°C \tag{2.31}$$

2.3.4 EC1 Parameter Fire Curve

In EC1 [3], the parameter fire model developed by Wickström is adopted for calculation in real fires. In the heating phase, the fire temperature is

$$T_g = 20 + 1{,}325(1 - 0.324e^{-0.2t^*} - 0.204e^{-1.7t^*} - 0.472e^{-19t^*}) \tag{2.32}$$

where, $t^* = t\Gamma$, in which

$$\Gamma = \frac{(F_o/\sqrt{k\rho c})^2}{(0.04/1{,}160)^2}. \tag{2.33}$$

$\sqrt{k\rho c}$ is the thermal inert of the enclosure.

The duration of the heating is determined by

$$t_{\max} = \max(\frac{0.2 \times 10^{-3}q_t}{F_o}, t_{lim}) \tag{2.34}$$

where, t_{lim} = 25, 20 and 15 min for slow, medium and fast fires, respectively. If $t_{\max} = 0.2 \times 10^{-3}q_t/F_o$, the fire is ventilation controlled; and if $t_{\max} = t_{lim}$, the fire is fuel controlled. For fuel controlled fires, when using Eq. 2.33 to calculate Γ_{lim}, $F_{o,lim}$ is taken as $0.1 \times 10^{-3}q_t/t_{lim}$.

Equation 2.32 is valid for

$$100 \leq \sqrt{k\rho c} \leq 2{,}200 \tag{2.35}$$

and

$$0.02 \leq F_o \leq 0.20 \tag{2.36}$$

In the cooling phases, the fire temperature is calculated by

$$T_g = T_{\max} - 625(t^* - t^*_{\max} \cdot x) \quad (t^*_{\max} \leq 0.5) \tag{2.37a}$$
$$T_g = T_{\max} - 250(3 - t^*_{\max})(t^* - t^*_{\max} \cdot x) \quad (0.5 < t^*_{\max} \leq 2) \tag{2.37b}$$
$$T_g = T_{\max} - 250(t^* - t^*_{\max} \cdot x) \quad (t^*_{\max} \geq 2) \tag{2.37c}$$

where, $t^*_{\max} = (0.2 \times 10^{-3}q_t/F_o)\Gamma$. For $t_{\max} > t_{lim}$, x=1; and for $t_{\max} = t_{lim}$, $x = t_{lim}\Gamma/t^*_{\max}$.

Lie correlations are only applicable for ventilation controlled fires, whilst EC1 correlations are applicable for both ventilation and fuel controlled fires. For ventilation controlled fires, the duration of the heating calculated by Lie and EC1 correlations are equal, thus

$$\tau = \frac{q_t A_t}{\dot{m}_{air}\Delta H_{c,air}} = \frac{q_t A_t}{0.52A_o\sqrt{H_o}\times 3.03 \times 3{,}600} \approx \frac{0.2\times 10^{-3}q_t}{F_o} \tag{2.38}$$

$$= \frac{17.5q_{t,wood}A_t}{0.52A_o\sqrt{H_o}\times 3.03 \times 3{,}600} \approx \frac{q_{t,wood}}{330F_o} \tag{2.39}$$

2.4 The Concept of Time Equivalent

2.4.1 Overview

The concept of equivalent fire severity, commonly referred as time equivalent, is used to relate the severity of an expected real fire to the standard fire. So that results or methods based on standard fire can be extended to realistic fires. The concept was originally developed by Ingberg [15]. Till now, a number of methods and empirical formulae have been developed for evaluating the equivalent fire severity. These methods include equal area method, maximum temperature method, minimum load capacity concept and energy based method [11, 16, 17].

Equal area method establishes time equivalency by equating the area under the time-temperature curve of design fire scenario to that under the standard fire scenario. If the areas under different curves are equal, the fires are considered to have equivalent severity. This method has no rational basis and might underestimate the heat transfer in a short hot fire and overestimate the heat transfer in a long cold fire, although both result in equal areas under two time-temperature curves [11].

Maximum temperature method defines the equivalent fire severity as the time of exposure to the standard fire that would result in the same maximum temperature in a protected steel member as would occur in a realistic fire. This method is the most widely used time equivalent concept. The empirical formulae for evaluating the equivalent fire severity, such as CIB, Law and Eurocode formulae [11], are based on this method. This method may not be accurate if the maximum temperatures used for computing the time equivalent are much higher or lower than those which would cause failure in a particular member.

Minimum load capacity method establishes the equivalent fire severity as the time of exposure to the standard fire that would result in the same load bearing capacity as the minimum which would occur in a realistic fire. This method is the most realistic time equivalent concept for the design of load bearing members, but it is difficult to implement for a material which does not have a clearly defined minimum load capacity.

The energy based method establishes the equivalency to the amount of energy transferred to the elements. Two fires will have the same fire severity if they transfer same amount of energy to an element. The method is originally proposed by Harmathy [16]. Harmathy [16] introduced a variable referred to as 'normalized heat load' to calculate the amount of the energy. The normalized heat load is defined as the total heat absorbed by a unit area of the boundaries of an enclosure during any fire, divided by the thermal absorptivity of the boundaries. The normalized heat load is the measure of the maximum temperature rise below the surface of a building element at a depth corresponding roughly to a distance where the important load-bearing components (e.g. steel bars) are located [17]. Recently, the energy based method has been adopted by Kodur et al. [17] to evaluate fire resistance of reinforced concrete beams.

2.4.2 EC1 Correlation

The Eurocode formula [3], based on maximum temperature method, is used to calculate the equivalent standard fire duration of natural fires. The formula is based on maximum temperature method which is derived for protected steel members.

Compare with other formulae based on maximum temperature method, the Eurocode formula is presented in a very simple form given by

$$t_{eq} = q_f k_b w_f \tag{2.40}$$

where, t_{eq} is the equivalent standard fire duration time in minutes; q_f is the floor fire load density; k_b is the conversion factor to account for the thermal properties of the enclosure; and w_f is the ventilation factor, for small compartments without horizontal vents

$$w_f = (\frac{A_o \sqrt{H_o}}{A_t})^{-1/2} \frac{A_f}{A_t} = F_o^{-1/2} \frac{A_f}{A_t} \tag{2.41}$$

2.5 Finite Element Simulation

2.5.1 Numerical Tool

The one-zone compartment fire model given above can be solved using technologies like finite differential method (FDM) and finite element method (FEM). In this section, the FEM program ANSYS is employed to simulate the model. ANSYS is power to solve steady-state or transient heat transfer problems, it capacity for using in fire environment has been validated by many works, e.g. [18, 19].

2.5.2 Basic Elements

2.5.2.1 LINK32-2D Conduction Bar

LINK32 is a uniaxial element with the ability to conduct heat between its nodes. The element has a single degree of freedom, temperature, at each node point. The conducting bar is applicable to a 2D, steady-state or transient thermal analysis.

The element is defined by two nodes, a cross-sectional area, and the material properties. The thermal conductivity is in the element longitudinal direction. Heat generation rates may be input as element body loads at the nodes.

2.5.2.2 LINK34-Convection Link

LINK34 is a uniaxial element with the ability to convect heat between its nodes. The element has a single degree of freedom, temperature, at each node point. The convection element is applicable to a 2D or 3D, steady-state or transient thermal analysis.

The element is defined by two nodes, a convection surface area, two empirical terms, and a film coefficient.

2.5.2.3 LINK31-Radiation Link

LINK31 is a uniaxial element which models the radiation heat flow rate between two points in space. The link has a single degree of freedom, temperature, at each node. The radiation element is applicable to a 2D or 3D, steady-state or transient thermal analysis.

The element is defined by two nodes, a radiating surface area, a geometric form factor, the emissivity, and the Stefan-Boltzmann constant.

2.5.2.4 MASS71-Thermal Mass

MASS71 is a point element having one degree of freedom, temperature, at the node. The element may be used in a transient thermal analysis to represent a body having thermal capacitance capability but negligible internal thermal resistance, that is, no significant temperature gradients within the body. The lumped thermal mass element is applicable to a 1-D, 2-D, or 3-D steady-state or transient thermal analysis.

The lumped thermal mass element is defined by one node and a thermal capacitance.

2.5.3 FE Model

Figure 2.4 illustrates the FE model. The heat source is modeled by a perfect conductor, which is represented by one LINK32 element. The heat generation rate of the source is input as *HRR* specified by NFSC. The compartment boundaries are modeled using LINK32. Convection and radiation at fire or environment exposed surfaces are

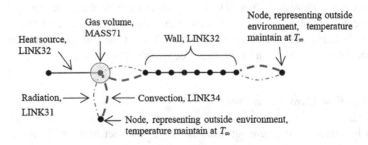

Fig. 2.4 FE Model for one zone compartment fire model

modeled using LINK34 and LINK31, respectively. Gas volume is modeled using MASS71. Radiation at opening is modeled using LINK31. Convection at opening by mass exchange is modeled using LINK34 with film coefficient of $c_p \dot{m}_{gas}/A_o$.

2.5.4 Verification of the FE Model

The results for fire temperatures predicted by the proposed model simulated by FEM are compared with those predicted by the program OZone [5], Lie correlations [14] and EC1 parameter fire model [3]. The program OZone V2 was developed by Cadorin et al. [5] to design structural steel elements submitted to compartment fires. In OZone V2, both two- and one-zone models are adopted to simulate the pre- and post-flashover fires. Under certain circumstances, the two-zone model can also switch automatically to a one-zone model.

In the study, $W \times D \times H = 3\,\text{m} \times 3\,\text{m} \times 2\,\text{m}$, $W_o \times H_o = 2\,\text{m} \times 1\,\text{m}$, $d_w = 0.2\,\text{m}$, $\rho_w = 2{,}300\,\text{kg/m}^3$, $c_w = 1{,}000\,\text{J/(kgK)}$, $k_w = 1.6\,\text{W/mK}$, $q_f = 800\,\text{MJ/m}^2$, $\alpha = 0.0117$, and the maximum heat release rate per unit area for fuel controlled fire is taken as $250\,\text{kW/m}^2$. Correspondingly, the calculated HRR_{max} for fuel and ventilation controlled fires are 2.25 and 3.0 MW respectively.

Figure 2.5 shows the comparison among the results of fire temperatures predicted by different methods. The FEM model and Ozone give consistent results for both fuel ($HRR_{\text{max}} = 2.25\,\text{MW}$) and ventilation ($HRR_{\text{max}} = 3\,\text{MW}$) controlled fires, whilst in the heating phases the fire temperatures predicted by the FEM model are slightly higher than those predicted by Ozone. The differences among the results by FEM model and Ozone are due to different mathematic technologies adopted by them in solving sub-models in one zone post-flashover fire model [5]. Results by Lie method agree well with those by FEM model using $HRR_{\text{max}} = 2.25\,\text{MW}$ in the heating phases. Lie method and EC1 parameter fire give same results for fuel and ventilation controlled fires.

Fig. 2.5 Comparison among results of fire temperatures predicted by different methods

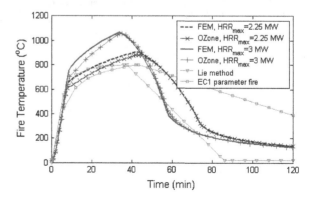

2.6 Conclusion

The widely used one-zone model for post-flashover fires has been presented. Based on the one-zone model, a FE model is developed to simulate the post-flashover fires. The FE model has been verified, which will be used in the following chapter.

References

1. D. Drysdale, *An Introduction to Fire Dynamics*, 2nd edn. (Wiley, New York, 1999)
2. J. Quintiere, *Fundamentals of Fire Phenomena* (Wiley, New York, 2006)
3. BSI, *Eurocode 1: Actions on Structures—Part 1–2: General Rules—Actions on Structures Exposed to Fire* (British Standard, UK, 2002)
4. V. Babrauskas, Compf2: a program for claculating post-flashover fire temperatures. Technical report, NBS Technical Note 991. National Bureau of Standards (1979)
5. J. Cadorin, J. Franssen, Fire Saf. J. **38**, 395 (2003)
6. W. Jones, R. Peacock, G. Forney, P. Reneke, Cfast—consolidated model of fire growth and smoke transport (version 6): Technical reference guide. Technical report, NIST Special Publication 1026, National Institude of Standard and Technology (NIST) (2009)
7. K. McGrattan, R. McDermott, S. Hostikka, J. Floyd, Fire dynamics simulator (version 5) user's guide. Technical report, NIST Special Publication 1019-5, NIST (2010)
8. A. Hasofer, V. Beck, Fire Saf. J. **28**, 207 (1997)
9. V. Bertola, E. Cafaro, Proc. R. Soc. A **465**, 1029 (2009)
10. O. Pettersson, S. Magnusson, J. Thor, *Fire Engineering Design of Steel Structures* (Swedish Institute of Steel Construction, Sweden, 1976)
11. A. Buchanan, *Structural Design for Fire Safety* (Wiley, New York, 2001)
12. V. Babrauskas, *SFPE Handbook of Fire Protection Engineering*, 3rd edn. (Society of Fire Protection Engineers, 2002), chap. Section 3–1: Heat Release Rates
13. B. McCaffrey, J. Quintiere, M. Harkleroad, Fire Technol. **17**(2), 98 (1981)
14. T. Lie, *SFPE Handbbok of Fire Protection Engineering*, 3rd edn. (Society of Fire Protection Engineers, 2002), chap. Section 4–8: Fire Temperature-Time Relations
15. S. Ingberg, NFPA Q. **22**, 43 (1928)
16. T. Harmathy, Fire and Mater. **11**, 95 (1987)
17. V. Kodur, P. Pakala, M. Dwaikat, Fire Saf. J. **45**, 211 (2010)
18. G. Li, C. Zhang, Int. J. Adv. Steel Constr. **6**(2), 788 (2010)
19. C. Zhang, G. Li, Y. Wang, Fire Technol. (2011)

Chapter 3
Steel Temperature in Natural Fires

3.1 Introduction

Post-flashover fires are widely adopted to represent natural fires. In literature, the temperature of a steel member exposed to a post-flashover fire is usually determined by first modeling the fire phenomenon by a empirical correlation (e.g. parameter fire model [1]) or advanced compute simulation (e.g. fire dynamic simulation [2]) to obtain a temperature-time curve to represent the fire environment, then substituting the fire curve into a 1D condensed heat transfer model to obtain the steel temperature [3]. The temperature of steel members in a fire can also be determined by advanced compute simulations [4].

At present, various formulae are provided by fire codes in different countries for calculating the temperature of insulated steel members in fire [3]. In deriving those formulae by different technologies like separation of variables, laplace transform and green's function approach, the standard fire curve is always adopted to represent the fire environment [3]. The current formulae, which are based on the standard fire, might give unacceptable results for calculation in natural fires [5].

When using the current formulae mentioned above, iterative computations should be always processed, which is not convenient for daily design works. In this chapter, a simple expression, which only needs hand calculation, has been proposed for calculating the maximum steel temperature of insulated steel members in natural fires.

3.2 One-Dimensional Heat Transfer Model

Figure 3.1 shows the one-dimensional (1D) heat transfer model used for calculating the temperature of steel members insulated by coatings [3, 5]. The temperature specified by a fire curve (T_f) is interpreted as the effective black body radiation temperature for radiation calculation and as the same gas temperature for convection calculation. Due to its high conductivity, the temperature gradient within the steel section has been ignored in the model.

© Springer-Verlag Berlin Heidelberg 2015
C. Zhang, *Reliability of Steel Columns Protected by Intumescent Coatings Subjected to Natural Fires*, Springer Theses, DOI 10.1007/978-3-662-46379-6_3

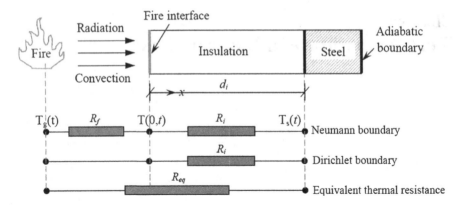

Fig. 3.1 1D condensed heat transfer model for predicting the temperature of insulated steel members

The governing heat transfer equation for the 1D model is given by

$$\alpha_i \frac{\partial^2 T(x, t)}{\partial x^2} - \frac{\partial T(x, t)}{\partial t} = 0 \tag{3.1}$$

where, $\alpha_i = k_i / c_i \rho_i$ is the thermal diffusivity; k_i is the thermal conductivity; and $c_i \rho_i$ is the volumetric specific heat of the insulation.

At the steel-insulation interface, the boundary condition is given by

$$- k_i \frac{\partial T(d_i, t)}{\partial x} = \frac{c_s \rho_s}{A_i / V} \frac{\partial T(d_i, t)}{\partial t} \tag{3.2}$$

$$T_s(t) = T(d_i, t) \tag{3.3}$$

where, $c_s \rho_s$ is the volumetric specific heat of the steel; A_i / V is the section factor, in which A_i is the appropriate area of the fire insulation material per unit length, and V is the volume of the steel per unit length; and d_i is the thickness of the insulation.

At the fire-insulation interface, two boundary conditions, namely Neumann and Dirichlet boundaries, have been used in engineering, which are given by

$$- k_i \frac{\partial T(0, t)}{\partial x} = \dot{q}_{in} \tag{3.4}$$

and

$$T(0, t) = T_g(t), \tag{3.5}$$

respectively. $T_g(t)$ and $T(0, t)$ are the temperatures of the fire and the insulation surface, respectively; $\dot{q}_{in}$ is the incident heat flux, for calculation in most fire conditions

$$\dot{q}_{in} = \dot{q}_c + \dot{q}_r = (h_c + h_r)[T_g(t) - T(0, t)] \tag{3.6}$$

where, h_c is the convection heat transfer coefficient, taken as 25 W/(m^{2}K) for nominal fires and 35 W/(m^{2}K) for natural fires [1]; and h_r is the radiative heat transfer coefficient, given by

$$h_r = \sigma \varepsilon_{res} [(T_g(t) + 273)^2$$
$$+ (T(0, t) + 273)^2] \times [T_g(t) + 273 + T(0, t) + 273] \qquad (3.7)$$

where, $\sigma = 5.67 \times 10^{-8}$ W/(m^{2}K^4) is the Stefan-Boltzmann constant; and ε_{res} is the resultant emissivity at the insulation surface, which is dependent on many parameters such as external heat flux, surface temperatures [6, 7]. In practice, constant value of ε_{res} is used, e.g. for intumescent coatings $\varepsilon_{res} = 0.92$ has been used in [6]. In standard fire, the surface temperature of the insulation, $T(0, t)$, is usually very close to the gas temperature that Eq. 3.7 is often simplified by [8].

$$h_r = 4\sigma \varepsilon_{res} [T_g(t) + 273]^3 \qquad (3.8)$$

3.3 Current Calculation Formulae

3.3.1 CECS200 Formula

In deriving the formula in the Chinese Code CECS200, the following assumptions have been adopted, that,

- The Dirichlet boundary is safely assumed at fire-insulation interface;
- The temperature distribution within the insulation is linear; and
- The temperature distribution within the steel is uniform.

At time increment Δt, the total energy transferred to the steel is

$$\Delta Q = \frac{T_g(t) - T_s(t)}{d_i / k_i} A_i \Delta t \qquad (3.9)$$

The energy absorbed by the steel is

$$\Delta Q_s = c_s \rho_s V \Delta T_s \qquad (3.10)$$

The energy absorbed by the insulation is

$$\Delta Q_i = \frac{\Delta T_s + \Delta T_g}{2} c_i \rho_i A_i d_i \qquad (3.11)$$

By energy balance, $\Delta Q = \Delta Q_s + \Delta Q_i$, we have

$$\Delta T_s = \frac{k_i A_i / V}{c_s \rho_s d_i} \frac{T_g - T_s}{(1 + \mu/2)} \Delta t - \frac{\Delta T_g}{2/\mu + 1} \qquad (3.12)$$

where

$$\mu = \frac{c_i \rho_i}{c_s \rho_s} d_i (A_i / V) \tag{3.13}$$

Ignoring the right second term in Eq. 3.12, we get the formula recommended by CECS200, that is

$$\Delta T_s = \frac{k_i A_i / V}{c_s \rho_s d_i} \frac{T_g - T_s}{(1 + \mu/2)} \Delta t \tag{3.14}$$

3.3.2 Pettersson et al.'s Formula

Assume a Neumann boundary at the fire-insulation interface and consider the effect of convection and radiation by adding a thermal resistance, $1/(h_c + h_r)$, to the denominator of the right term of Eq. 3.9, Pettersson et al. [9] proposed a formula which has been widely used in fire engineering, that

$$\Delta T_s = \frac{A_i / V}{c_s \rho_s} \frac{1}{(\frac{d_i}{k_i} + \frac{1}{h_c + h_r})} \frac{T_g - T_s}{1 + \mu/2} \Delta t - \frac{\Delta T_g}{2/\mu + 1} \tag{3.15}$$

3.3.3 EC3 Formula

The formula provided by EC3 [10] was originally derived by Wickstrom [11], through solving the 1D heat transfer model with a Dirichlet boundary at the fire-insulation interface. Separation of variable technology was adopted in the derivation. The formula is given by

$$\Delta T_s = \frac{k_i A_i / V}{c_s \rho_s d_i} \frac{T_g - T_s}{1 + \mu/3} \Delta t - (e^{\mu/10} - 1) \Delta T_g \tag{3.16}$$

3.3.4 ECCS Formula

The formula provided by ECCS [12] is derived by using laplace transfer technology to solve the 1D heat transfer model with a Neumann boundary at the fire-insulation interface [13], which is given by

$$\frac{dT_s}{dt} = A'(T_g - T_s) - B' \frac{T_g}{dt} \tag{3.17}$$

where

$$A' = \frac{1}{(\frac{c_s \rho_s}{A_i/V})(\frac{d_i}{k_i} + \frac{1}{h_c+hr})(1 + \frac{\mu}{N})} \tag{3.18}$$

$$B' = \frac{b}{1 + N/\mu} \tag{3.19}$$

with N and b as weighting factors. Certainly, for limiting case $(h_c + h_r) \to \infty$ the Neumann boundary is equivalent to the Dirichlet boundary. At this case,

$$b = \frac{1 + \mu/4}{2(1 + 5\mu/8)} \tag{3.20}$$

$$N = 2(b + 1) \tag{3.21}$$

3.3.5 Silva's Formula

Assume a Dirichlet boundary, by an analytical process, Silva [14] derived a formula which had been recommended for the revision of the Brazilian Standard 14323, that

$$\Delta T_s = \frac{k_i A_i/V}{c_s \rho_s d_i} \frac{T_g - T_s}{1 + \mu/4} \Delta t - \frac{\Delta T_g}{4/\mu + 1} \tag{3.22}$$

3.4 Shortcomings of the Current Formulae

The current formulae mentioned above are originally developed for calculating in standard fire which only includes heating phase. In [5], the applicability of using those formulae for calculation in natural fires which include heating and cooling phases had been investigated. The widely referenced 'Swedish' fire curves and measured temperature time curves in real fire tests were adopted to represent different natural fire environments. Parameters including insulation thickness, section factor, and protection material were considered in sensitivity studies. The steel temperatures predicted by different formulae were compared with the numerical results predicted by FEM. Comparatively, the formula given by Chinese Code CECS200 gives best prediction of steel temperatures.

The current formulae are developed as "simplified" methods, that their applications are limited to situations where the properties of the insulation materials are or can be treated as constant or temperature-independent. Besides, when calculating using the current formulae mentioned above, iterative computations should be always processed, which is not convenient for daily design works.

3.5 Proposed Simple Approach

3.5.1 Methodology

When considering the structural behavior of a steel member in fire, it is usually assumed that the load capacity of the member is only related to the the maximum temperature it reached that the effect of heating rate is ignorable (for composite structures like composite slab, however, the effect of heating rate should be considered [15]). In other words, if the maximum temperatures of a same steel member in different fires are equal, the load capacities of the member in those fires are consistent. Correspondingly, the approaches for calculating the load capacity of steel members, which are based on standard fire tests conducted in laboratory, are applied for practical design of the fire resistance of steel members exposed to potential real fires. As a result, with considering the complexities in both modeling real fires and simulating the structural behavior in real fires (which usually includes advanced compute simulations), if the severity of a real fire can be represented by an equivalent duration in the standard fire, it will simplify the daily design work greatly. Thus, the concept of time equivalent had been developed to relate real fires with the standard fire. Detail description of the concept can be found in Sect. 2.4 in Chap. 2.

The simple closed-form expression, given by ECCS [16], to calculate the temperature of insulated steel members exposed to the standard fire is given by

$$T_s = \frac{t}{40}(\frac{A_i/V}{d_i/k_i})^{0.77} + 140 \tag{3.23}$$

where, t is the standard fire exposure time. Equation 3.23 was originally obtained from curve fitting of test data and was developed for calculating the limiting temperature of steel members in standard fire [16]. The limiting temperature is the temperature when steel members fail, which usually ranges from 400 to 600 °C, as shown in Fig. 3.2.

Fig. 3.2 Calculated steel temperatures by Eqs. 3.14 and 3.23 (The *black bold lines* are steel temperatures calculated by Eq. 3.14)

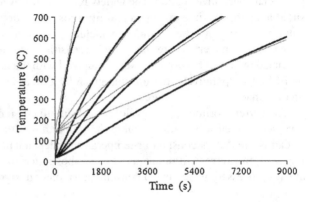

Calculating the equivalent standard fire exposure time of a natural fire by using Eurocode formula given by Eq. 2.40, and taking the calculated time into Eq. 3.23, we get

$$T_{smax0} = \frac{q_f k_b w_f}{40} (\frac{A_i / V}{d_i / k_i})^{0.77} + 140 \qquad (3.24)$$

3.5.2 Compare with OZone V2

3.5.2.1 The Program OZone V2

The program OZone V2 [17] was adopted to evaluate the accuracy of using Eq. 3.24 to calculate the maximum temperature of insulated steel members to natural fires. Ventilation controlled post-flashover fires are considered and one-zone model is used in the simulations. The natural fire safety concept (NFSC) [18] is used to represent the fire conditions. In OZone V2, the temperature of insulated steel members to fire is calculated by step-to-step calculations using the equation given in EC3 [10].

It should be noted that the equation given in EC3 gives negative temperature increments at the early heating stage which is invalid from a physical point of view. To avoid this, in practice, the steel temperature increments are taken as zero if the values calculated by the equation are negative [10]. Such treatment, however, might cause significant over-prediction of the maximum steel temperatures in natural fires [5]. With considering the fact that in OZone V2 the above treatment is adopted, the predicted maximum steel temperatures by OZone V2 might be much higher than the real values.

3.5.2.2 Investigated Cases

Table 3.1 gives values of input parameters in the case studies. Totally 698 cases are included. Three typical fireproofing materials are considered, which are light weight cementitious or mineral fiber spray-applied fire resistive material (SFRM), intumescent coating (IC) and gypsum board (Gypsum). The constant thermal properties of the SFRM ($\rho_i = 300$ kg/m^3, $c_i = 1,200$ J/kgK, and $k_i = 0.12$ W/mK) and the gypsum board ($\rho_i = 800$ kg/m^3, $c_i = 1,700$ J/kgK, and $k_i = 0.2$ W/mK) given in OZone V2 are used. The equivalent constant thermal conductivity of the intumescent coating ($k_i = 0.04$ W/mK) is taken from [19]. Different insulation thicknesses, which are corresponding to different fire resistance ratings, are considered. The values in the bracket are the corresponding fire resistance ratings. Different natural fires in a compartment with dimensions of 3 m width, 4 m length and 2.7 m height are considered. The compartment has a vent with height of 2 m. The width of the vent is varied to simulate different fire scenario. Also, the fire load density is varied to produce different fires. Three typical section factors are investigated, which are 75, 200 and 300. The section factors investigated cover the range of commonly used

Table 3.1 Investigated cases in OZone V2

Materials	Cases	d_i (mm)	A_i/V (m^{-1})	q_f (MJ/m^2)	F_o (m$^{1/2}$)
SFRM	20	4.17 (0.5 h)	200	240–600	0.046
SFRM	51	10.27 (1.0 h)	200	240–1,120	0.046
SFRM	39	17.39 (1.5 h)	200	240–1,720	0.046
SFRM	39	25.26 (2.0 h)	200	240–1,720	0.046
SFRM	39	33.76 (2.5 h)	200	240–1,720	0.046
SFRM	39	42.78 (3.0 h)	200	480–2,700	0.046
SFRM	39	17.39 (1.5 h)	75	400–2,250	0.046
SFRM	26	17.39 (1.5 h)	300	240–1,200	0.046
SFRM	35	17.39 (1.5 h)	200	750	0.023–0.201
IC	17	1.39 (0.5 h)	200	120–600	0.046
IC	34	5.80 (1.5 h)	200	240–1,520	0.046
IC	51	14.26 (3.0 h)	200	240–2,200	0.046
IC	34	5.80 (1.5 h)	75	600–1,880	0.046
IC	26	5.80 (1.5 h)	300	240–1,200	0.046
IC	41	3.42 (1.0 h)	200	750	0.023–0.201
Gypsum	14	6.96 (0.5 h)	200	120–600	0.046
Gypsum	39	28.98 (1.5 h)	200	240–1,720	0.046
Gypsum	24	71.29 (3.0 h)	200	400–1,500	0.046
Gypsum	38	28.98 (1.5 h)	75	600–2,400	0.046
Gypsum	18	28.98 (1.5 h)	300	200–1,000	0.046
Gypsum	35	17.12 (1.0 h)	200	750	0.023–0.174

steel sections in engineering. The opening factor, $F_o = A_o\sqrt{H_o}/A_t$ in Table 3.1, is presented to show different vent conditions (different vent width).

3.5.2.3 Results and Discussions

Figure 3.3a shows the comparison among the results for maximum temperature of steel members protected by SFRM, predicted by OZone V2 and Eq. 3.24. On the whole, Eq. 3.24 diverges from OZone V2 significantly, and gives lower maximum steel temperatures. By data fitting using Matlab, the maximum steel temperatures predicted by OZone V2 can be best represented by the following quadratic equation, thus

$$T_{smax} = -0.0024T_{smax0}^2 + 3.2T_{smax0} - 400. \quad (3.25)$$

Take

$$\Delta = \frac{q_f k_b w_f}{40}\left(\frac{A_i/V}{d_i/\lambda_i}\right)^{0.77} \quad (3.26)$$

Fig. 3.3 Comparison among the maximum temperatures of insulated steel members, predicted by OZone V2 and Eq. 3.24. **a** SFRM, **b** intumescent coating, **c** gypsum board

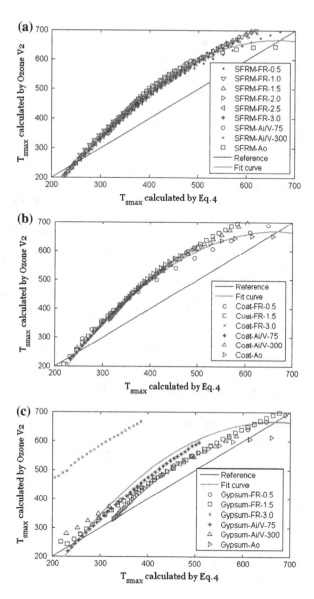

and consider Eqs. 3.24, 3.25 can also be expressed as

$$T_{s\max} = -0.0024\Delta^2 + 2.528\Delta + 0.96 \tag{3.27}$$

The fitted curve is also shown in Fig. 3.3a. In the range of $T_{s\max}$ from 300 to 600 °C, the differences among the results predicted by OZone V2 and Eqs. 3.25 or 3.27 are small (within 20 °C).

Figure 3.3b shows the comparison among the results for maximum temperature of steel members protected by intumescent coating, predicted by OZone V2 and Eq. 3.24. The maximum temperatures calculated by Eq. 3.25 are also presented. In the range of T_{smax} from 300 to 600 °C, the differences among the results predicted by OZone V2 and Eq. 3.25 are small (within 20 °C).

Figure 3.3c shows the comparison among the results for maximum temperature of steel members protected by gypsum board, predicted by OZone V2 and Eq. 3.24. The maximum temperatures calculated by Eq. 3.25 are also presented. For all cases except cases with insulation thickness of 71.29 mm, in the range of T_{smax} from 300 to 600 °C, comparing with OZone V2, Eq. 3.25 gives acceptable predictions (the maximum over-prediction by Eq. 3.25 reaches about 60 °C).

As being mentioned before, the treatment adopted by EC3 equation to avoid negative temperature increment results in over-prediction of maximum steel temperatures in natural fires. Figure 3.4 shows this over-prediction [5]. When deriving EC3 formula by separation of variables, a Dirichlet boundary is assumed at the fire exposed surface [3]. The assumption gives good prediction of steel temperatures for insulation with low density and low conductivity (e.g. SFRMs and intumecent coatings), but yields conservative results for insulation with high density and high conductivity (e.g. gypsum board and NWC) [3]. The treatment and the assumption together leads to significant over-prediction of maximum steel temperatures for cases with insulation of 71.29 mm gypsum board shown in Fig. 3.3c.

3.5.2.4 The Modified One Zone Model

It is obvious that the steel members in a fire compartment will absorb a portion of the heat released by combustion. That portion of heat will heat the steel members on one hand and cool the compartment on the other hand. As a result, the temperature of a steel member within a fire compartment is dependent on the heating mechanism of the compartment. However, in current model as mentioned before, the temperature of a steel member within a fire compartment is related to the fire curve which is

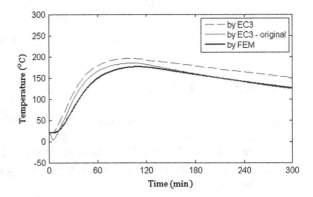

Fig. 3.4 Over-prediction of maximum steel temperature in a natural fire caused by the treatment adopted by EC3 [5] ('EC3-original' corresponds to the equation without adopting the treatment)

determined without considering the heat sink effect of the steel member. Here, the heat sink effect of steel members in fire compartments has been considered by adding a quantity to the traditional heat balance equation for one zone compartment model (Eq. 2.1 in Sect. 2.2.1), thus

$$HRR = \dot{q}_g + \dot{q}_w + \dot{q}_{o,c} + \dot{q}_{o,r} + \dot{q}_s \tag{3.28}$$

where $\dot{q}_s$ is rate of heat storage in steel members.

3.5.3 Compare with FEM

Solve Eq. 3.28, we can obtain both gas and steel temperatures in fire compartments. Here, the verified FEM model in Sect. 2.5.3 is modified to consider the heating sink effect of steel members. Figure 3.5 shows the modified FEM model.

3.5.3.1 Monte Carlo Simulations

Using the above FEM model, Monte Carlo Simulations are conducted to obtain data for comparison. In each simulation loop, the maximum steel temperatures predicted by Eqs. 3.24, 3.27 and FEM are recorded. Table 3.2 gives the input variables for Monte Carlo Simulations. Totally, three cases (marked by using 'A', 'B' and 'C' in Table 3.2) are investigated, where each case includes 200 simulation loops.

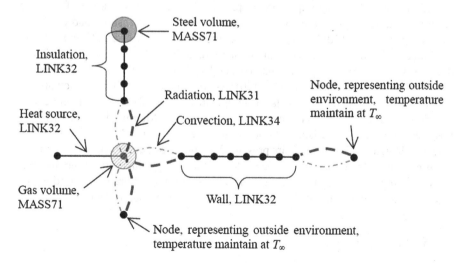

Fig. 3.5 Illustration of the FEM thermal model for heat transfer analysis

Table 3.2 Input variables for Monte Carlo Simulations

Variable	Mean	COV	Distribution
q_f	600 MJ/m^2	0.4	Lognormal
A_f	12 m^2	–	Deterministic
A_t	61.8 m^2	–	Deterministic
A_v	2 m^2	0.2	Normal
ρ_w	2,300 kg/m^3	–	Deterministic
c_w	1,000 J/kgK	–	Deterministic
k_w	1.6 W/mK	0.1	Normal
d_w	200 mm	0.1	Normal
A_i/V	190 m^{-1}	–	Deterministic
k_i	0.04^A, 0.12^B, and 0.2^C W/mK	0.3	Normal
d_i	2^A, 9^B, and 10^C mm	0.2	Lognormal

Compare with FEM, Eq. 3.24 fails to give acceptable prediction of maximum steel temperatures. Figure 3.6a–c give the comparisons among the results for the maximum steel temperatures predicted by Eq. 3.27 and FEM, which show acceptable match in the range of temperatures from 300 to 600 °C. On the whole, Eq. 3.27 gives higher maximum steel temperatures and the over-prediction in most cases are within 50 °C, as shown in Fig. 3.7.

3.5.4 Compare with Test Data

Test data on temperature of steel members in natural fires are limited. Konicek and Lie [20] reported tests on temperature of protected steel columns under different fire severities. Table 3.3 gives the parameters for the tests. The measured and predicted steel temperatures in [20], along with the results calculated by Eqs. 3.24 and 3.27 are presented in the table. Compare with measured data, Eq. 3.24 gives better prediction than Eq. 3.27. However, Eq. 3.27 matches good with predicted data. The differences are within 50 °C. The predicted data in [20] were obtained from compute calculations.

Kirby et al. [21] reported tests on temperature of insulated steel members to natural fires. Table 3.4 gives the parameters for the tests. Measured data in different positions in each test, marked by "Back", "Middle" and "Front", along with results calculated by Eqs. 3.24 and 3.27 are presented. In the report, thermal properties of fire proofing material were not given. The value of thermal conductivity of the material derived from the reported temperature-time curve in standard fire test is used in our calculations. Equation 3.27 matches very well with the measured data.

3.5.5 Profession Factor

Professional factor is used to account for model error in calculation approaches, which is defined as the ratio of measured and predicted results.

Fig. 3.6 Comparison among the maximum temperatures of insulated steel members, predicted by OZone V2 and Eq. 3.27. **a** Case A, **b** case B, **c** case C

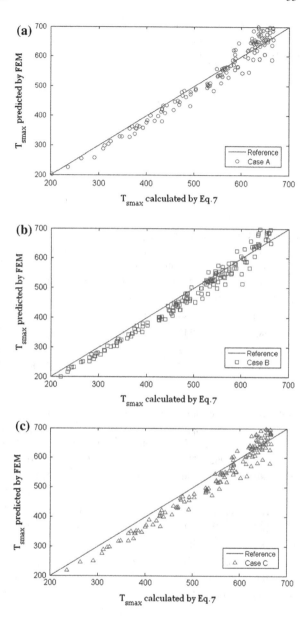

The professional factor or model error for Eq. 3.27 is characterized by using the above test data, which has a mean of 0.955 and a COV of 0.014, and can be best described by lognormal distribution as shown in Fig. 3.8. The probability density function for the lognormal distribution is given by

Fig. 3.7 Differences among
the results for maximum
steel temperatures predicted
by Eq. 3.27 and FEM

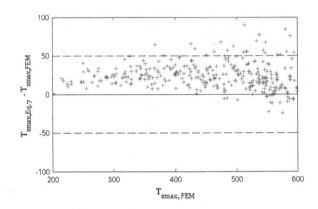

Table 3.3 Compare with test data given by Konicek and Lie [20]

Test	Ai/V (m^{-1})	O $(m^{1/2})$	q_f (MJ/m^2)	k_i $(W/(mK))$	d_i (mm)	Measured T_{smax}	T_{smax} Predicted	Eq. 3.24	Eq. 3.27
1	110	0.05	288.8	0.187	25	460.9	508.4	447.8	551.7
2	110	0.05	350.3	0.187	25	517.4	561.4	513.3	610.2
3	110	0.05	384.5	0.187	25	587.8	633.3	549.8	633.9
4	58.3	0.02	347.7	0.187	25	498.9	551.1	499.6	599.7
5	38.1	0.1	867.2	0.187	25	489.6	517.8	429.0	531.1
6	38.1	0.1	577.6	0.187	25	374.8	366.5	332.5	398.6

Table 3.4 Compare with test data given by Kirby et al. [21]

Test	Ai/V (m^{-1})	O $(m^{1/2})$	q_f (MJ/m^2)	k_b $(min \cdot m^2/MJ)$	d_i (mm)	Back	T_{smax} Middle	Front	Eq. 3.24	Eq. 3.27
1	180	0.062	759.9	0.07	30	588.0	–	616.5	451.9	556.0
2	180	0.062	380.1	0.07	30	378.0	428.0	308.0	296.0	336.9
3	180	0.031	380.1	0.07	30	400.5	442.0	418.0	382.6	473.0
4	180	0.031	759.9	0.07	30	616.5	653.0	678.5	625.0	662.5
5	180	0.016	380.1	0.07	30	493.0	521.5	538.0	447.2	551.1
6	180	0.008	380.1	0.07	30	605	590	559.5	486.0	588.4
7	180	0.051	380.1	0.07	30	–	–	258.0	293.4	332.2
8	180	0.060	402.3	0.07	30	361.0	382.0	330.0	309.4	360.4
9	180	0.062	380.1	0.07	30	379.0	403.5	308.5	296.0	336.9

$$f(x) = \frac{1}{x\sigma\sqrt{2\pi}} e^{\frac{-(\ln x - \mu)^2}{2\sigma^2}} \tag{3.29}$$

where $\mu = -0.053455$ and $\sigma = 0.121189$ are parameters.

Fig. 3.8 Probabilistic
property for professional
factor of Eq. 3.27. **a**
Probability density, **b**
cumulative probability, **c**
probability plot

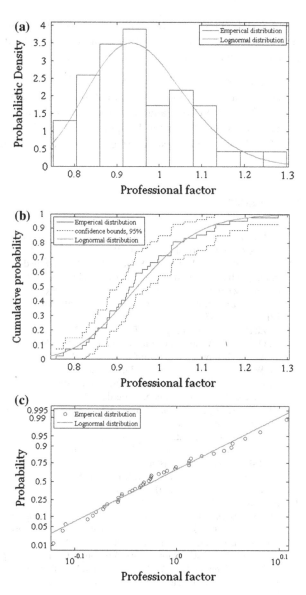

3.6 Conclusions

In different codes, various formulae are provided for calculating the temperature
of insulated steel members in fire. However, those formulae are based on standard
fire and might give unacceptable results for calculation in natural fires. A simple

approach has been proposed to calculate the maximum steel temperature of insulated steel members in natural fires. Based on the results mentioned above, the following conclusions can be drawn,

- The proposed approach can give acceptable prediction of the maximum steel temperature of insulated members in natural fires. The proposed equation, Eqs. 3.25 or 3.27, is valid in the range of maximum steel temperatures from 300 to 600 °C. The professional factor for the approach Eq. 3.27 has a mean of 0.955 and a COV of 0.014, and can be best described lognormal distribution.

References

1. BSI, *Eurocode 1: Actions on Structures—Part 1–2: General Rules—Actions on Structures Exposed to Fire* (British Standard, UK, 2002)
2. K. McGrattan, R. McDermott, S. Hostikka, J. Floyd, Fire dynamics simulator (version 5) user's guide. Technical Report, NIST. NIST Special Publication 1019-5 (2010)
3. G. Li, C. Zhang, Int. J. Adv. Steel Constr. **6**(2), 788 (2010)
4. G. Li, C. Zhang, in *Proceedings of the Sixth International Conference on Structures in Fire*, MI, 2010, pp. 35–42
5. C. Zhang, G. Li, Y. Wang, Fire Technol. (2011)
6. A. Omrane, Y. Wang, U. Goransson, G. Holmstedt, M. Aldn. Fire Saf. J. **42**, 68 (2007)
7. J. Staggs, H. Phylaktou, Fire Saf. J. **43**, 1 (2008)
8. M. Wong, J. Ghojel, Fire Saf. J. **38**, 187 (2003)
9. O. Pettersson, S. Magnusson, J. Thor, *Fire Engineering Design of Steel Structures* (Swedish Institute of Steel Construction, Rotterdam, 1976)
10. BSI, *Eurocode 3: Design of Steel Structures—Part 1–2—General Rules—Structural Fire Design* (British Standard, UK, 2005)
11. U. Wickstrom, Fire Saf. J. **9**, 281 (1985)
12. ETC 3, *European Recommendations for the Fire Safety of Steel Structures* (Elsevier Scientific Publishing Company, Amsterdam, 1983)
13. S. Melinek, P. Thomas, Fire Saf. J. **12**, 1 (1987)
14. V. Silva, Eng. Struct. **27**, 2036 (2005)
15. S. Lamont, A. Usmani, M. Gillie, Fire Saf. J. **39**, 327 (2004)
16. ECCS, *Design Manual on the European Recommendation for the Fire Safety of Steel Structures* (European Commission for Constructional Steelwork, Brussels, 1985)
17. J. Cadorin, J. Franssen, Fire Saf. J. **38**, 395 (2003)
18. E. Coal, S.C. (ECSC), Natural fire safety concept, valorization project. Technical Report (2001)
19. G. Li, C. Zhang, G. Lou, Y. Wang, L. Wang, Fire Technol. (2011)
20. L. Konicek, T. Lie, Fire tests on protected steel columns under different fire severities. Technical Report (1974)
21. B. Kirby, D. Wainman, L. Tomlinson, T. Kay, B. Peacock, Natural fires in large scale compartments—a British steel technical, fire research station collaborative project. Technical Report (British Steel, 1994)

Chapter 4
Thermal Properties of Intumescent Coatings in Fire

4.1 Introduction

Intumescent coatings, by their advantages like attractive appearance, potential for off-site application and practically taking no space, are now the dominant passive fire protection materials used in industrial and public buildings [1]. The coatings, which usually are composed of inorganic components contained in a polymer matrix, are inert at low temperatures and will expand and degrade to provide a charred layer of low conductivity materials at temperatures of approximately 280–350 °C [2, 3]. The charred layer, which acts as thermal barrier, will prevent heat transfer to underlying substrate.

In current codes, the fire resistance of a coating is measured using standard fire tests for rating the materials [4, 5]. In such tests, a large steel member (in Chinese code, the tested sample is a 0.5 m length steel I beam [6]) is coated with the fireproofing material then inserted in a furnace that is heated following the standard temperature–time curve. The time for the steel member to exceed any of the endpoint failure criteria confers the rating of the coating. The widely used endpoint failure criteria is that the maximum mean steel temperature must be lower than the critical temperature which is the temperature that causes structure collapse in fire situation, often taken as 550 °C. Such tests are expensive and time-consuming, with a large number of tests required to cover the range of steel configurations and protection thicknesses typically required in construction.

Alternatively, if thermal resistance of the coating can be derived, calculation methods are available and efficient to assess the fire resistance. Unlike the conventional fireproofing materials (e.g. concrete, gypsum, SFRMs) whose thermal properties are temperature-dependent only, the performance of intumescent coatings are complex that they will behave differently according to the applied heating condition, coating thickness, and protected structures [7–10]. As a result, the traditional standard test methods (like ASTM C518-04 [11], GB/T 10294-2008 [12]) are not applicable to measure the thermal properties of intumescent coatings [13].

© Springer-Verlag Berlin Heidelberg 2015
C. Zhang, *Reliability of Steel Columns Protected by Intumescent Coatings Subjected to Natural Fires*, Springer Theses, DOI 10.1007/978-3-662-46379-6_4

Till now, several models have been developed to study the heat transfer of intumescent coatings under heating [2, 7, 13–15]. These models are primarily one-dimensional, and concentrated on the effects of swelling on the thermal properties of coatings. The structure of chars is always simply assumed to be constituted of vapor and solid materials, or to be a porous media, and the thermal conductivity of chars, k_c, is determined by k_{sol}, k_{vap} and f_v [2, 13, 15] (in Refs. [7, 14], the effect of thermal radiation in bubbles on k_c have also been considered). Here, k_c, k_{sol} and k_{vap} are thermal conductivities of char, solid and vapor, respectively; and f_v is the void fraction of the char. In intumescing or swelling process, the structure of coatings is divided into two layers, virgin coating and char [7, 13, 15] (in Refs. [9, 14], a transforming swelling layer is included between the virgin and char layers). In [8], the thermal conductivity of a commercial intumescent coating is measured using the relationship between thermal conductivity and thermal diffusion, where thermal diffusivity is measured by a designed laser flash diffusivity system.

Due to the complexity of intumescing process and the difficulty of measuring the structure of chars, the thermal conductivity of intumescent coatings can not be measured directly. In fire engineering, effective thermal conductivity or equivalent thermal resistance is usually adopted to characterize the thermal insulation property of intumescent coatings. Anderson et al. [13] developed a procedure to estimate the effective thermal conductivity of chars of intumescent systems. The procedure was based on a heat transfer analysis of temperature–time data from one-dimensionally designed experiments of coated coupons exposed to a fire environment typical of aviation-type fuel fires. Bartholmai et al. [10] developed a simple test method to determine the time dependent thermal conductivity of intumescent coatings. The method consists of temperature measurements using the bench-scaled experimental set-up of a cone calorimeter and finite difference simulation to calculate the effective thermal conductivity. The simulation procedure was also adapted to the small scale test furnace, in which the standard temperature–time curve was applied to a larger sample and thus which provided results relevant for approval. In DD ENV 13381-4:2002 [16], the inverse equation of the EC3 [17] equation for calculating the temperature of insulated steel members to fire is presented to extract the effective thermal conductivity of intumescent coatings. Dai et al. [18] used the inverse equation for calculating the temperatures in steel joints with partially protected by intumescent coatings, which gives acceptable results.

When using the procedures mentioned above to calculate the effective thermal conductivity of intumescent coatings, complex compute simulations or iterative calculation procedures are usually required which is not convenient for daily design work. In structural fire safety design, the limiting temperature (instead of the whole heating history) of key elements is concerned by the designer, and acceptable simple formulae have been developed for calculating the limiting temperature of protected steel members in standard fire, e.g. ECCS [19], CECS [20]. Those simple formulae are, however, only applicable to situations where the properties of the insulation materials are or can be treated as constant or temperature-independent [21].

In this chapter, a simple procedure has been developed to determine the equivalent constant thermal resistance of intumescent coatings.

4.2 Theoretical Models for Intumescent Coatings

4.2.1 Intumescing Process

The intumescent coatings are usually composed of a combination of an acid source (ammonium phosphate, APP), a carbon source (pentaerythritol, PER) and a blowing agent (melamine). These ingredients are bound together by a polymer matrix. When exposed to flame or radiation, the coatings expand and regrade to provide an insulating, formed char surface over the underlying substrate. The char is of low reactivity and provides an impermeable barrier of high thermal resistance.

As shown in Fig. 4.1, when exposed to flame or radiation, broadly, a intumescent coating undergoes the following reaction steps [3, 14, 15],

- At early heating stage, a large amount of thermal energy is absorbed by the coating, whose temperature increases quickly.
- When the temperature of the coating reaches a critical temperature, the polymer matrix melts and degrades to form a viscous fluid. The inorganic acid source in the coating will undergo thermal decomposition normally at temperature of 100–250 °C [3].
- At temperature of 280–350 °C [2, 3], the bowing agent within the coating decomposes to release a large amounts of gas of which some fraction is trapped within the molten matrix.
- The molten fluid hardens and releases residual volatile to form char.

4.2.2 Thermal Conductivity of the Char

Firstly, it is assumed that the structure of intumesced char is constituted solely of vapor and solid material, and that the cell size of the pores is sufficiently small that convective currents are suppressed, and that thermal radiation does not have a "direct"

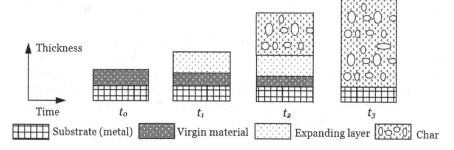

Fig. 4.1 Illustration of the intumescing process

path through the char to the substrate. Then it is assumed that the arrangement of the solid material and vapor, integrated through the thickness of the char, can be considered a thermal resistance network. The thermal conductivity of the char can be computed as follows [13],

$$\frac{1}{k_c} = \frac{1 - f_v}{k_{sol}} + \frac{f_v}{k_{vap}} \qquad (4.1)$$

4.3 Equivalent Thermal Resistance of Intumescent Coatings

4.3.1 Equivalent Thermal Resistance

Figure 3.1 also shows the thermal resistance networks for the 1D condensed models using Neumann and Dirichlet boundaries, in which,

$$R_i = \frac{d_i}{k_i} \qquad (4.2)$$

and

$$R_f = \frac{1}{h_c + h_r}. \qquad (4.3)$$

Here, R_i, R_f are the thermal resistance of the insulation, and the thermal resistance caused by convection and radiation.

The Neumann boundary is complex and has capacity to represent the real boundary condition at the fire interface. From Eq. 3.7, we know to get h_r, the value of the surface temperature of the insulation, $T(0, t)$, should be known beforehand. However, $T(0, t)$ is a unknown variable, and for intumescent coatings in fire the measurement of $T(0, t)$ is very difficult [22]. The Dirichlet boundary is simple which assumes $T(0, t)$ is equal to the surrounding gas temperature $T_g(t)$. The Dirichlet boundary ignores the heat loss through surface convection and radiation. This assumption is valid for conditions where $R_i \gg R_f$, but will yield conservative results for conditions where insulation is not effective.

Instead of calculating R_f and R_i directly and separately, using an equivalent thermal resistor R_{eq}, as shown in Fig. 3.1, can represent all thermal energy blocking effects caused by convection, radiation, and insulation. For most calculations using the Dirichlet boundary condition where $R_f = 0$, the equivalent thermal resistance R_{eq} is equal to R_i. The equivalent thermal resistance is efficient and useful to evaluate the fire resistance of intumescent coatings.

Rearrange Eq. 3.14, we get the equation for calculating the equivalent thermal resistance of coatings that

$$R_{eq} = \frac{T_g(t) - T_s(t)}{\Delta T_s / \Delta t} \frac{1}{c_s \rho_s} \frac{A_i}{V} \tag{4.4}$$

From the equivalent thermal resistance, we can easily get the effective thermal conductivity of the intumescent coatings by

$$k_{eff} = \frac{d_{io}}{R_{eq}} \tag{4.5}$$

here, d_{io} is the initial thickness of the coatings. In DD ENV13381-4:2002 [16], the following equation is provided for calculating the effective thermal conductivity of the coatings, thus

$$k_{eff}Z = [\frac{d_{io}}{A_i/V} \times c_s \rho_s \times (1 + \mu/3) \times \frac{1}{(T_g(t) - T_s(t))\Delta t}]$$
$$\times [\Delta T_s + (e^{\mu/10} - 1)\Delta T_s]. \tag{4.6}$$

For intumescent coatings, the mass ratio ϕ can be approximated as zero, and Eq. 4.6 becomes Eq. 4.5.

4.3.2 Equivalent Constant Thermal Resistance

Rearranging Eq. 3.23, we get the expression to calculate the equivalent constant thermal resistance, thus

$$R_{const} = \frac{d_i}{k_i} = (\frac{t}{2,400(T_{crit} - 140)})^{1/0.77} \frac{A_i}{V} \tag{4.7}$$

where T_{crit} is the critical temperature, which usually ranges from 400 to 600 °C.

4.4 Experimental Investigation

4.4.1 Test on Small-Scale Samples

4.4.1.1 Test Approach

A small scale test furnace has been constructed for fire resistance testing. Figure 4.2 is a picture of the furnace. The dimensions of the firebox are 1.0 m (length) × 1.0 m (width) × 1.2 m (height). Heating system is computer controlled, which has capacity of simulating standard ISO834 fire, standard hydrocarbon fire and user-defined fires. Figure 4.3 gives the comparison between the measured furnace temperature–time curve and the standard ISO834 fire curve, which shows good match.

Fig. 4.2 A picture of the
furnace used in test on
small-scale samples

Fig. 4.3 Measured furnace
fire curve and the ISO834
standard fire curve

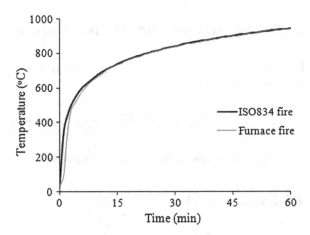

Steel plates with two small holes are used as the test samples (substrates).
Figure 4.4 shows the dimensions of the plate and the applying of the insulation.
The section factor of the plate is taken as $A_i/V = 125$ m^{-1}. In tests, samples are
hinged on the supports through holes in the samples, as shown in Fig. 4.5. The sample
is designed to represent the 1D heat transfer model discussed above.

Temperatures are measured at two arbitrary measuring points located on the steel
plate surface (there are four measuring points on one sample) using type-K Nicr-Ni
thermocouple. The average value of the measured temperatures is taken as the steel
plate temperature.

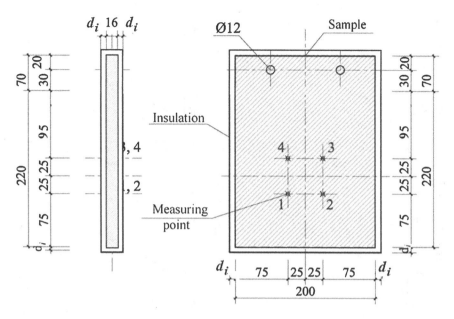

Fig. 4.4 Dimensions of the small-scale sample

Fig. 4.5 Small-scale
samples hinged in furnace

4.4.1.2 Test Data

In another research project [23], which focused on investigating the effects of aging
on thermal properties of intumescent coating for steel elements, the test approach
mentioned above was used for testing. In the project, accelerated aging and fire tests

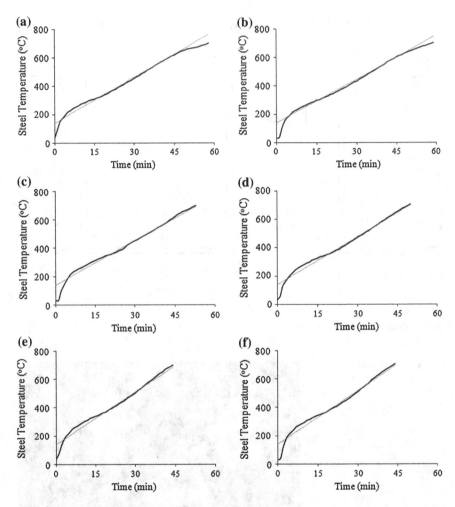

Fig. 4.6 Measured steel temperatures for test 1 on specimens with 1 mm intumescent coating in [23] (where 0 to 42 are numbers of cycles in accelerated aging tests; and the *black bold lines* are measured steel temperatures). **a** 0 cycle, **b** 2 cycles, **c** 4 cycles, **d** 11 cycles, **e** 21 cycles, **f** 42 cycles

were conducted on 36 specimens, 18 of which were applied with 1 mm coating and the other 18 with 2 mm coating. Accelerated aging tests were conducted by according to the European Code ETAG 018-2 [24]. The equation given by DD ENV13381-4:2002 [16], or Eq. 4.6, was used to calculate the effective thermal conductivity of the coatings. Figures 4.6 and 4.7 give the results for measured steel temperatures, in which 21 cycles of accelerated aging was assumed to represent working life of 10 years (and 42 cycles, 20 years, etc.).

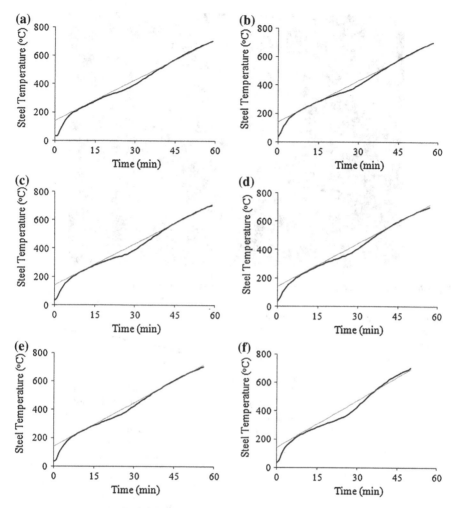

Fig. 4.7 Measured steel temperatures for test 2 on specimens with 2 mm intumescent coating in [23] (where 0 to 42 are numbers of cycles in accelerated aging tests; and the *black bold lines* are measured steel temperatures). **a** 0 cycle, **b** 2 cycles, **c** 4 cycles, **d** 11 cycles, **e** 21 cycles, **f** 42 cycles. **a** 0 cycle, **b** 2 cycles, **c** 4 cycles, **d** 11 cycles, **e** 21 cycles, **f** 42 cycles

4.4.2 Test on Steel Members

4.4.2.1 Test Approach

Dai et al. [18] tested the temperatures in steel joints with partial intumescent coating fire protection exposed to the standard fire. The furnace at the University of Manchester was used for testing. Figure 4.8 shows a exterior view of the furnace. The internal dimensions of the furnace are 3.5 m × 3 m × 2.5 m. Figure 4.9 shows the furnace temperatures were close to the ISO834 fire.

Fig. 4.8 A exterior view of the furnace used in test on steel members [18]

Fig. 4.9 Measured furnace temperatures and the ISO834 fire [18]

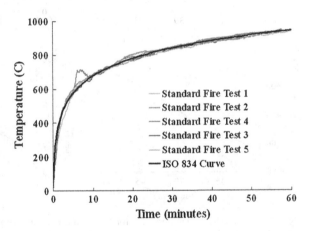

In their tests, each steel assembly consisted of one column and four beams, which were connected together by bolts, as shown in Fig. 4.10. The column was 250 × 254 × 89 UC, and with length of 1,000 mm. All of the four beams had the same sections 305 × 165 × 40 UB. Numerous thermocouples were used to monitor the temperature distributions at different locations of the steel sections.

4.4.2.2 Test Data

In Ref. [18], totally 10 tests on joints with different fire-protection schemes using intumescent coating were conducted. Intumescent coating fire protection was applied

Fig. 4.10 A picture of tested steel assembles in [3, 18]

Fig. 4.11 Steel column temperatures in [3, 18] (the *black bold lines* are measured steel temperatures)

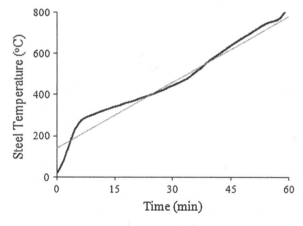

by the intumescent coating manufacturers' own application team. Figure 4.11 gives the results for measured temperatures in test 1 in Ref. [18]. The average coat thickness for the column in test 1 is 0.67 mm.

4.4.3 T_s Calculated by Using R_{const}

In Figs. 4.6, 4.7 and 4.11, the steel temperatures calculated by Eq. 3.23 are also presented, which match well with the test data in the range of steel temperatures from 400 to 600 °C. In those calculations, constant thermal resistance determined by Eq. 4.7 with $T_{crit} = 550$ °C (and the corresponding t_{crit} which is the time when the measured steel temperature reaches $T_{crit} = 550$ °C) are used.

Figure 4.12 shows some results for equivalent thermal resistance, R_{eq}, calculated by using Eq. 4.7 with replacing T_{crit} by the measured steel temperatures, for small-scale tests. Figure 4.13 shows the result for equivalent thermal resistance for the full-scale test. At low temperatures, the calculated R_{eq} change greatly with temperature increase; whilst at high temperatures, R_{eq} almost maintain at constant values. This is because at low temperatures, the intumescent coatings react and swell that the structure and property of the coating system change greatly but at high temperatures, reaction of the intumescent coatings has finished and the final inert charred structure has been formed, as illustrated in Fig. 4.1. The equivalent constant thermal resistance, R_{const}, used in calculations, are also plotted in Figs. 4.12 and 4.13. Table 4.1 gives the values of R_{const} obtained from small-scale tests.

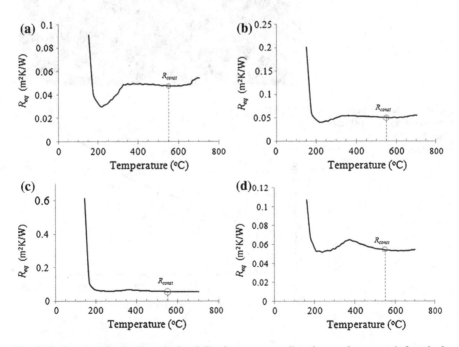

Fig. 4.12 Some results for the calculated R_{eq} for test on small-scale samples. **a** test 1, 0 cycle, **b** test 1, 2 cycles, **c** test 2, 0 cycles, **d** test 2, 2 cycles

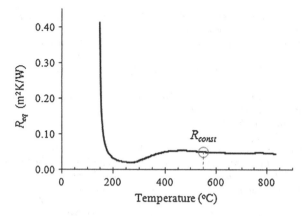

Fig. 4.13 Calculated R_{eq} for test on steel members ($R_{const} = 0.0500 \ \text{Km}^2/\text{W}$)

Table 4.1 Constant thermal resistance obtained from small-scale tests (units in Km^2/W)

	0 cycles	2 cycles	4 cycles	11 cycles	21 cycles	42 cycles
Test 1	0.0514	0.0504	0.0491	0.0448	0.0396	0.0385
Test 2	0.0552	0.0542	0.0522	0.0516	0.0505	0.0458

4.5 Conclusions

The behavior of intumescent coatings under heating is very complex and no agreeable model is available to simulate the behavior. Effective thermal conductivity or equivalent thermal resistance is usually used to evaluate the fire resistance of intumescent coatings. However, complex compute simulations are always required to predict the time/temperature-dependent effective thermal conductivity. This chapter proposed a simple procedure to assess the fire resistance of intumescent coatings by using equivalent constant thermal resistance. The main conclusion is

- The procedure is valid and convenient to assess the fire resistance of intumescent coatings. Using the equivalent constant thermal resistance of intumescent coatings determined by the procedure, the calculated steel temperatures agree well with the test data in the range of the concerned limiting temperatures from 400 to 600 °C.

References

1. Corus, *Fire Resistance of Steel-Framed Buildings*, 2006 edn. (Corus Construction and Industrial, Scunthorpe, 2006)
2. S. Bourbigot, S. Duquesne, J. Leroy, J. Fire Sci. **17**, 42 (1999)
3. J. Yuan, Intumescent coating performance on steel structures under realistic fire conditions. Ph.D. thesis, The University of Manchester (2009)

4. A. E119-00a, *Standard Test Methods for Fire Tests of Building Construction and Materials* (ASTM International, West Conshohocken, 2000)
5. G. 9978-1999, *Fire-Resistance Tests—Elements of Building Construction* (Chinese National Code, Canada,1999)
6. G. 14907-2002, *Fire Resistive Coating for Steel Structure* (Chinese National Code, Canada, 2002)
7. J. Yuan, Y. Wang, in *Proceedings of the Fifth International Conference on Structures in Fire (SiF'08)*, Singapore, 2008, pp. 713–724
8. J. Koo, Fire Technol. **34**, 59 (1998)
9. C.D. Blasi, J. Anal. Appl. Pyrol. **71**, 721 (2004)
10. M. Bartholmai, R. Schriever, B. Schartel, Fire Mater. **27**, 151 (2003)
11. A. C518-04, *Standard Test Method for Steady-State Thermal Transmission Properties by Means of the Heat Flow Meter Apparatus* (ASTM International, West Conshohocken, 2003)
12. G. 10294-2008, *Thermal Insulation—Determination of Steady-State Thermal Resistance and Related Properties—Guarded Hot Plate Apparatus* (Chinese National Code, Canada, 2008)
13. C. Anderson, D. Ketchum, W. Mountain, J. Fire Sci. **6**, 390 (1988)
14. G. Griffin, J. Fire Sci. **00**, 1 (2009)
15. M. Gillet, L. Autrique, L. Perez, J. Phys. D Appl. Phys. **40**, 883 (2007)
16. D.E. 13381-4, *Test Methods for Determining the Contribution to the Fire Resistance of Structural Members—Part 4: Applied Protection to Steel Members* (British Standards Institution, London, 2002)
17. BSI, *Eurocode 3: Design of Steel Structures—Part 1–2: General Rules—Structural Fire Design* (British Standard, 2005)
18. X. Dai, Y. Wang, C. Bailey, Fire Technol. **46**, 19 (2010)
19. ECCS, *Design Manual on the European Recommendation for the Fire Safety of Steel Structures* (European Commission for Constructional Steelwork, Brussels, 1985)
20. C.A. for Engineering Construction Standardization (CECS200), *Technical Code for Fire Safety of Steel Structure in Buildings* (China Planning Press, Beijing, 2006)
21. C. Zhang, G. Li, Y. Wang, Fire Technol. **48**(2), 343–366 (2011)
22. A. Omrane, Y. Wang, U. Goransson, G. Holmstedt, M. Aldn, Fire Saf. J. **42**, 68 (2007)
23. L. Wang, Y. Wang, G. Li, in *Proceedings of the Sixth International Conference on Structures in Fire*, MI, 2010, pp. 735–742
24. ETAG018, *Guideline for European Technical Approval of Fire Protective Products, Part 2: Reactive Coatings for Fire Protection of Steel Elements* (European Organization for Technical Approvals, 2006)

Chapter 5
Behavior of Steel Columns in Fire

5.1 Introduction

Columns are the critical building components, failure of which usually leads to progressive collapse of the local or global structures. Fire induced high temperature has two aspects of effects on a steel column, which are on one hand it reduces the strength and stiffness of the steel material and on the other hand it causes expansion of the column. If a non-uniform temperature distribution forms through the section, thermal curvature will occur. Any resistance to the free movement of axial thermal expansion or thermal curvature will induce internal stresses within the member.

In prescriptive codes, to ensure structural fire safety, the fire resistance of a steel column should be not less than the rating specified in building regulations. The fire resistance of a building component is defined as the time when the component exceeds any of the endpoint failure criteria in standard fire tests. In BS476-20 [1], the failure of steel columns occurs when the test isolated member fails to support the test load. Due to its high conductivity, bare steel is frangible to fire. As a result, steel columns always need fire protection to achieve the specified fire resistance ratings.

As being mentioned in Chap. 2, the behavior of a real fire is complex, which depends on many parameters such as active fire detection and suppression systems (smoke detector and sprinkler), fire load (amount and distribution), combustion, ventilation, compartment size and geometry, and thermal properties of compartment boundaries [2]. In many situations (e.g. large enclosures with limited fire load or where sprinklers work effectively), the severities of potential fires are much lower than that of the standard fire. Besides, several real fire accidents and tests (e.g. Broadgate fire and Cardington full-scale fire tests [3]) have shown that global structures are more robust in fire than simply-supported independent structural components and the fire-resistance capacity of components coupled in global structures are normally greater than that of isolated components. Considering those two main shortcomings, which are standard fire bears little resemblance to a real fire and the behavior of tested isolated members cannot represent the global behavior of structures in fire condition, the prescriptive approaches usually yield conservative design.

© Springer-Verlag Berlin Heidelberg 2015
C. Zhang, *Reliability of Steel Columns Protected by Intumescent Coatings Subjected to Natural Fires*, Springer Theses, DOI 10.1007/978-3-662-46379-6_5

As an alternative to the traditional prescriptive approaches, performance-based (PB) approaches had been prompted worldwide to do optimum fire resistance design. To develop PB approaches, in the past decades, many researchers had studied the robust behavior of steel columns in fire conditions experimentally [4–8] and numerically [9–12]. Also, simple calculation approaches had been developed for predicting the critical temperature of steel columns [13–15].

When using calculation approaches for fire resistance design, the user should be awake of the accuracy and limitations of the approaches being adopted. Also, in probabilistic analysis, the model error or professional factor of the deterministic approach should be determined. This chapter presents a comparative study on critical temperatures of steel columns. The accuracy and limitations of calculation approaches for predicting the buckling and limit temperature of steel column are investigated by comparing with test data reported in literature. The probabilistic property of professional factor of calculation approaches are also characterized by the test data.

5.2 Fundamental Principles

When subjected to fire, the total strain of steel is composed of [16]

$$\varepsilon = \varepsilon_{mec}(\sigma, T) + \varepsilon_{th}(T) + \varepsilon_{cr}(\sigma, T, t) \tag{5.1}$$

where $\varepsilon_{th} = \varepsilon_T + \varepsilon_\phi$ is thermal strain, in which ε_T is strain caused by thermal expansion due to uniform temperature rise and ε_ϕ is strain caused by thermal bowing due to temperature gradient in the section. For steel columns in fire, the temperature distribution within the section is always assumed to be uniform. ε_{mec} is mechanical strain; and ε_{cr} is creep strain, which is usually not considered in fire resistance analysis.

As shown in Fig. 5.1a, strain caused by thermal expansion due to a uniform temperature rise, ΔT, is given by

$$\varepsilon_T = \alpha \Delta T \tag{5.2}$$

where α is coefficient of thermal expansion. For case where axial displacement of the member is totally restrained (Fig. 5.1b), from Eq. 5.1 = 0 we get $\varepsilon_\sigma = -\varepsilon_T$ and the restraining force is

$$\Delta P = E_T A \varepsilon_\sigma = -E_T A \varepsilon_T = -E_T A \alpha \Delta T. \tag{5.3}$$

For a sufficient stocky member, with temperature rise the member will sooner or later fail due to yield of the material. The yield temperature increment is given by

$$\Delta T_y = \frac{f_{yT}}{\alpha E_T} \tag{5.4}$$

Fig. 5.1 Effect of thermal expansion on unloaded member

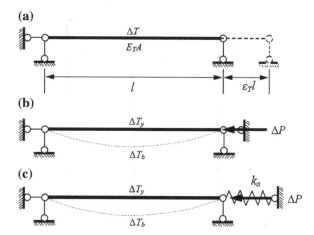

where f_{yT} and E_T are high temperature yield stress and high temperature elastic modulus, respectively. For a slender member, with temperature rise the member will fail because of buckling. Take Euler buckling load $P_{E,T} = \pi^2 E_T I / l^2$ into Eq. 5.3, we get the buckling temperature increment,

$$\Delta T_b = \frac{\pi^2}{\alpha}(\frac{r}{l})^2 = \frac{\pi^2}{\alpha\lambda^2} \qquad (5.5)$$

where, λ is the slenderness ratio of the member.

For case where axial displacement of the member is finitely restrained (Fig. 5.1c), assume stiffness of the axial restraint is k_a, the restraining force is calculated by

$$\Delta P = \frac{E_T A\alpha\Delta T}{1 + E_T A/(k_a l)}. \qquad (5.6)$$

The elastic buckling temperature increment is given by

$$\Delta T_b = \frac{\pi^2}{\alpha\lambda^2}(1 + \frac{E_T A}{k_a l}) \qquad (5.7)$$

As shown in Fig. 5.2, for case where a finitely restrained column is supporting a initial axial force P_0 before heating, the restraining force ΔP due to uniform temperature rise can be calculated as follows [17]

$$\Delta l = \frac{P_0}{k_c} - \frac{P_0}{k_{c0}} + \frac{\Delta P}{k_c} = \Delta\varepsilon_{mec}l + \frac{\Delta P}{k_c} \qquad (5.8)$$

and

$$\Delta = \varepsilon_{th}l - \Delta l = \frac{\Delta P}{k_a}. \qquad (5.9)$$

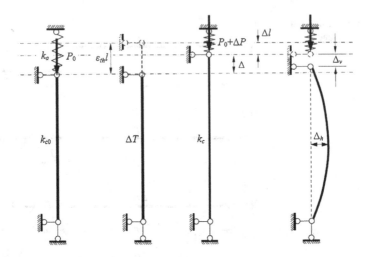

Fig. 5.2 Effect of thermal expansion on an axially loaded restrained column

Then

$$\Delta P = \frac{k_c k_a}{k_c + k_a}(\varepsilon_{th} - \Delta\varepsilon_{mec})l \qquad (5.10)$$

where

$$\Delta\varepsilon_{mec} = \frac{P_0}{k_c l} - \frac{P_0}{k_{c0} l}. \qquad (5.11)$$

k_{c0}, k_c are the axial stiffness of the column at room and high temperatures, respectively.

Equation 5.10 is valid for calculation in pre-buckling situations. For post-buckling situations, the axial contraction, Δ_v in Fig. 5.2, caused by lateral movement of the column should be considered, and the restraining force is given by [10]:

$$\Delta P = \frac{k_c k_a}{k_c + k_a}(\varepsilon_{th}l - \Delta\varepsilon_{mec}l - \Delta_v) \qquad (5.12)$$

5.3 Design Method

5.3.1 Failure Criterion

Structural components are designed to fulfill the function of supporting loads acting on them. Correspondingly, the limit state for a component is the state at which its load-bearing capacity equals to value of the design load. If a component losses its capacity to support the design load, from the point of view of a designer, the component fails.

Fig. 5.3 Load-temperature relationship of an axially restrained steel column subjected to fire

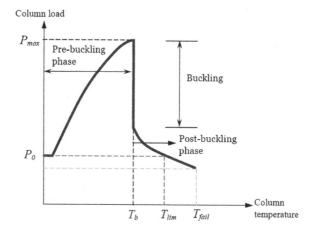

When subjected to fire, due to deterioration of material properties at high temperature, the load-capacity of structures decreases with temperature rise. For a free column, during heating process the acting force on the column maintains constant as the value of the design load at room temperature. The column buckles when temperature reaches so high at which the buckling capacity of the column decreases to the value of the acting or design force. After buckling, the free column will be no longer to support the design force. The limit state for free columns in fire is the state when buckling occurs, and the limit (failure) temperature of free columns is the buckling temperature.

For an axially restrained steel column in fire, from Eq. 5.10 we know that the value of the acting force on steel column changes with temperature rise. Figure 5.3 shows the complete load-temperature relationship of an axially restrained steel column subjected to fire. At the pre-buckling phase, the acting force on the column or the column load increases with temperature rise till buckling occurs at which the column load reaches its maximum value P_{max}. After buckling, the restraint helps the column to support load and the column load decreases with temperature rise. If the restraint has sufficient stiffness, the column load will decrease to values lower than the initial design load P_0. According to definition of the limit state given above, the column has definitely failed when column load reaches the initial design load, and the column can no longer support further temperature rise. However, in real situations, a restrained column can continue to support temperature rise till the restraint fails. T_b, T_{lim}, and T_{fail} in Fig. 5.3 are the buckling, limit and (actual) failure temperatures of restrained steel columns, respectively.

5.3.2 Free Steel Column in Fire

The simple formula developed by Franssen et al. [13, 18] is adopted by EC3 [19] for calculating the buckling resistance of axially loaded steel column in fire, which is given by

$$N_{b,T} = \chi_T A f_{yT} \qquad\qquad (5.13)$$

where,

$$\chi_T = \frac{1}{\varphi_T + \sqrt{\varphi_T{}^2 - \bar{\lambda}_T{}^2}} \qquad\qquad (5.14)$$

$$\varphi_T = \frac{1}{2}[1 + \alpha\bar{\lambda}_T + \bar{\lambda}_T{}^2] \qquad\qquad (5.15)$$

$$\alpha = 0.65\sqrt{\frac{235}{f_{y20}}} \qquad\qquad (5.16)$$

$$\bar{\lambda}_T = \bar{\lambda}_{20}\sqrt{\frac{k_{y,T}}{k_{E,T}}} = \sqrt{\frac{A f_{yT}}{P_{E,T}}}. \qquad\qquad (5.17)$$

Here, 0.65 is the severity factor at fire condition.

From Eq. 5.13, we can get the buckling temperature of steel column by solving

$$\frac{P_T}{N_{b,T}} = 1 \qquad\qquad (5.18)$$

with

$$P_T = P_0 + \Delta P \qquad\qquad (5.19)$$

where P_T is the column service load at fire condition. For free column in fire, $\Delta P = 0$.

In EC3 [19], a formula has been presented for calculating the buckling temperature of steel column in fire, that

$$T_b = 39.19\ln(\frac{1}{0.9674\mu_0{}^{3.833}} - 1) + 482 \qquad\qquad (5.20)$$

where $\mu_0 = \frac{P_T}{N_{b,0}}$ is utilization factor, where $N_{b,0}$ is the column buckling resistance at room temperature. μ_0 must not be taken less than 0.013.

5.3.3 Restrained Steel Column in Fire

Before a restrained column buckles, it can be designed as an axially loaded column. The buckling temperature of axially restrained steel column can also be derived from Eq. 5.18 with calculating ΔP by Eq. 5.10.

The limit temperature of restrained steel column in fire can be calculated follow the steps given by Wang [10]. Based on the calculation using the finite element program FINEFIRE, Neves et al. [14] proposed a simplified method by modifying the failure temperature of free column, that

$$T_{lim} = T_b^{free} - C_b \Delta T_{lim} \tag{5.21}$$

where C_b is a parameter, which takes the value of 0.9 when the column bends around the major axis and 1.25 when the column bends around the minor axis; ΔT_{lim} is the reduction in limit temperature caused by the axial restraint,

$$\Delta T_{lim} = \begin{cases} \frac{\Delta T}{0.03} \beta_l & \text{if } \beta_l \leq 0.03 \\ \\ \Delta T & \text{if } \beta_l > 0.03 \end{cases}$$

in which $\beta_l = k_a/k_{c0}$ is the axial restraint stiffness ratio, and ΔT is a parameter given by

$$\Delta T = \begin{cases} 0 & \text{if } \lambda \leq 20 \\ \\ 85 C_\rho \frac{\lambda - 20}{20} & \text{if } 20 < \lambda \leq 40 \\ \\ [85 + \frac{140}{40}(\lambda - 40)] C_\rho & \text{if } 40 < \lambda \leq 80 \\ \\ (260 - 0.44\lambda) C_\rho & \text{if } 80 < \lambda \leq 200 \end{cases}$$

where $C_\rho = 0.3 + \mu_{20}$, is a parameter considering the effect of initial axial load ratio $\mu_{20} = P_0/N_{b,0}$ ($0.3 \leq \mu_{20} \leq 0.7$); λ is the column slenderness.

5.3.4 FEM Model

Figure 5.4 shows a FEM structural model of axially restrained columns. The steel column is modeled using 3D linear finite strain beam element, BEAM188. BEAM188 is based on Timoshenko beam theory and is suitable for analyzing slender to moderately stubby/thick beam structures. The axial restraint is modeled by an axial spring using spring-damper element, COMBIN14.

Fig. 5.4 Illustration of the FEM model of axially restrained steel column

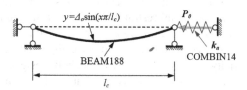

The initial column crookedness is assumed to be a half sine wave. The initial deflection amplitude at mid-height, if not specified, is taken as 0.1 % of the column length. Perfect elastic-plastic constitutive model is assumed. Residual stress is not considered. For free columns, the stiffness of the spring is taken as zero. For columns with other type of end conditions, the model can be modified by restraining the corresponding degrees of freedom to represent the conditions. Reduction factors for elastic modulus and yield stress of steel at high temperatures specified in EC3 [19] are used.

5.4 Test Data

5.4.1 Buckling Temperature of Free Column

When developing the EC3 formula for calculating the buckling resistance of axially loaded steel column in fire, Franssen et al. [13] summarized test data on free steel columns at high temperature reported in literature. Here, 69 tests summarized in [13] are considered . Table 5.1 gives the parameters for those tests. The test labels in [13] are used in Table 5.1. The ends of the column in those tests include pinned-pinned, pinned-fixed and fixed-fixed. In some tests, values of the yield stress of steel measured at flange and web were reported. In our calculations, uniform yield stress within the steel cross section is assumed and the values measured at flange in those tests are used. In Table 5.1, l for those tests are the effective lengths of the column.

Lie and Macaulay [20] reported 3 tests on fire resistance of protected steel columns. The test columns were 3,810 mm long. End conditions were fixed-fixed. The furnace heating condition followed ASTM-E119. The tests are labels as "Lie1" to "Lie3" in Table 5.1.

Ali et al. [5] reported 37 high temperature tests on steel columns with different degrees of axial restraint subjected to various load ratios. The test columns were 1,800 mm long with pinned ends. Tests on steel columns with zero degree of axial restraint are given in Table 5.1, which are labeled as "Ali1" to "Ali9". In [5], values of the initial load, P_0, were not given, instead values of the load ratio or utilization factor, μ_0, were presented. The load ratio in [5] was defined according to BS5950. The values of P_0 in Table 5.1 are calculated according to the definition in [5], but μ_0 in Table 5.1 are calculated according to EC3 [19] as in Eq. 5.20. In [5], material tests had not been conducted. Correspondingly, nominal values of material properties are used in Table 5.1.

Tan et al. [7] reported 15 tests on structural responses of restrained steel columns at elevated temperatures. The columns with different slenderness ratios were subjected to various axial restraint ratios. The test columns were 1,500 mm long with pinned ends. The columns were axially loaded and exposed to a monotonically increasing heating condition. Initial imperfections such as column crookedness and load eccentricity were measures by a specially designed facility. Tests on steel columns with no axial restraint are given in Table 5.1, which are labeled as "RS45-1" to "RS97-1".

Table 5.1 Measured and predicted buckling temperature for free columns

Label	$b_s \times h_s \times t_w \times t_f$ (mm×mm×mm×mm)	f_{y20} (Mpa)	e_{20} (Mpa)	l (mm)	e (mm)	Ends	Axis	λ_{20}	N_{b20} (kN)	P_0 (kN)	μ_0	T_b Meas	T_b FEM	Anal	EC3
7	300 × 300 × 11 × 19	271	2.05e5	1,890	0	F^a-F	W	24.4	3,270	2,000	0.61	588	581	549	551
8	82 × 160 × 5 × 7.4	271	2.05e5	1,890	0	F-F	W	100.8	208	104	0.50	564	600	554	584
9	82 × 160 × 5 × 7.4	271	2.05e5	1,890	0	F-F	W	100.8	208	151	0.73	475	537	451	518
11	100 × 200 × 5.6 × 8.5	277	2.05e5	1,890	0	F-F	W	82.8	365	266	0.73	394	554	468	518
12	100 × 200 × 5.6 × 8.5	271	2.05e5	1,915	0	F-F	W	83.9	355	324	0.91	250	509	269	453
13	120 × 120 × 6.5 × 11	260	2.05e5	1,890	0	F-F	W	60.8	531	363	0.68	519	585	511	530
14	120 × 120 × 6.5 × 11	260	2.05e5	1,890	0	F-F	W	60.8	531	267	0.50	585	638	569	584
15	180 × 180 × 8.5 × 14	275	2.05e5	1,890	0	F-F	W	40.8	1,296	603	0.46	603	646	590	596
16	180 × 180 × 8.5 × 14	275	2.05e5	1,890	0	F-F	W	40.8	1,296	893	0.69	560	574	518	529
17	190 × 200 × 6.5 × 10	279	2.05e5	1,890	0	F-F	W	39.4	1,043	677	0.65	556	584	531	540
18	290 × 300 × 8.5 × 14	269	2.05e5	1,890	0	F-F	W	25.6	2,351	1,507	0.64	561	572	539	543
19	210 × 220 × 7 × 11	252	2.05e5	1,915	0	F-F	W	36.0	1,175	972	0.83	502	530	461	487
20	200 × 200 × 9 × 15	218	2.05e5	1,915	0	F-F	W	37.1	1,268	681	0.54	549	614	569	573
21	140 × 140 × 7 × 12	247	2.05e5	1,915	0	F-F	W	52.8	696	542	0.78	516	558	473	503
22	140 × 140 × 7 × 12	247	2.05e5	1,915	0	F-F	W	52.8	696	372	0.53	576	625	563	574
23	110 × 220 × 5.9 × 9.2	273	2.05e5	1,915	0	F-F	W	75.9	461	319	0.69	522	571	500	528
31	120 × 120 × 6.5 × 11	257	2.05e5	3,800	0	P^b-P	S	75.2	453	318	0.70	560	584	496	525
33	220 × 220 × 9.5 × 16	261	2.05e5	3,800	12	P-P	S	40.3	1,726	767	0.44	590	622	597	603
34	210 × 220 × 7 × 11	309	2.05e5	3,800	12	P-P	S	39.7	1,386	784	0.57	560	581	558	564
36	120 × 120 × 6.5 × 11	257	2.05e5	4,800	12	P-P	S	94.9	364	105	0.29	685	685	638	670

(continued)

Table 5.1 (continued)

Label	$b_s \times h_s \times t_w \times t_f$ (mm×mm×mm×mm)	f_{y20} (Mpa)	e_{20} (Mpa)	l (mm)	e (mm)	Ends	Axis	λ_{20}	N_{b20} (kN)	P_0 (kN)	μ_0	T_b			
												Meas	FEM	Anal	EC3
54	180 × 180 × 8.5 × 14	267	2.05e5	3,860	0	P-P	S	50.3	1,160	891	0.77	475	563	480	506
55	180 × 180 × 8.5 × 14	250	2.05e5	3,860	0	P-P	S	50.3	1,092	876	0.80	446	554	462	495
56	220 × 220 × 9.5 × 16	269	2.05e5	3,700	12	P-P	S	39.2	1,792	1,489	0.83	335	494	456	486
57	220 × 220 × 9.5 × 16	250	2.05e5	3,700	12	P-P	S	39.2	1,671	1,628	0.97	230	413	245	405
58	220 × 220 × 9.5 × 16	263	2.05e5	3,700	12	P-P	S	39.2	1,754	1,136	0.65	478	555	532	541
60	160 × 160 × 8 × 13	262	2.05e5	4,700	12	P-P	S	69.2	783	755	0.96	232	444	192	416
61	160 × 160 × 8 × 13	259	2.05e5	4,700	12	P-P	S	69.2	775	590	0.76	412	523	465	508
62	160 × 160 × 8 × 13	249	2.05e5	4,700	12	P-P	S	69.2	751	650	0.87	290	492	401	473
78	240 × 240 × 10 × 17	229	2.05e5	3,700	0	P-P	W	59.8	1,496	1230	0.82	425	550	443	489
79	240 × 240 × 10 × 17	221	2.05e5	1,850	0	P-P	W	29.9	1,840	1,195	0.65	547	575	535	540
86	170 × 360 × 8 × 12.7	287	2.05e5	3,700	9	P-P	W	95.9	822	768	0.93	290	322	207	440
87	170 × 360 × 8 × 12.7	273	2.05e5	3,700	9	P-P	W	95.9	797	610	0.77	355	474	428	507
88	150 × 300 × 7.1 × 10.7	248	2.05e5	4,700	11	P-P	W	137.9	351	320	0.91	255	295	220	453
89	150 × 300 × 7.1 × 10.7	246	2.05e5	4,700	11	P-P	W	137.9	350	280	0.80	316	403	359	496
90	91 × 180 × 5.3 × 8	433	2.05e5	4,700	25	P-P	W	225.9	80	50	0.63	541	444	485	546
94	180 × 180 × 8.5 × 14	259	2.05e5	2,702	10	F-P	S	35.2	1,280	847	0.66	515	546	529	537
95	180 × 180 × 8.5 × 14	244	2.05e5	2,702	10	F-P	S	35.2	1,208	857	0.71	466	530	514	523
96	180 × 180 × 8.5 × 14	283	2.05e5	1,930	0	F-F	S	25.1	1,505	928	0.62	513	579	547	550
97	180 × 180 × 8.5 × 14	278	2.05e5	1,930	0	F-F	S	25.1	1,479	928	0.63	540	575	543	546
98	180 × 180 × 8.5 × 14	277	2.05e5	1,930	0	F-F	S	25.1	1,474	938	0.64	523	575	543	544

(continued)

Table 5.1 (continued)

Label	$b_s \times h_s \times t_w \times t_f$ (mm×mm×mm×mm)	f_{y20} (Mpa)	e_{20} (Mpa)	l (mm)	e (mm)	Ends	Axis	λ_{20}	N_{b20} (kN)	P_0 (kN)	μ_0	T_b			
												Meas	FEM	Anal	EC3
103	100 × 96 × 5 × 8	300	2.05e5	1,994	0	P-P	W	77.2	303	337	1.11	365	444	21	–
104	100 × 96 × 5 × 8	300	2.05e5	1,994	0	P-P	W	77.2	303	318	1.05	400	476	21	–
105	100 × 96 × 5 × 8	300	2.05e5	1,994	0	P-P	W	77.2	303	250	0.82	510	533	405	488
106	100 × 96 × 5 × 8	300	2.05e5	1,994	0	P-P	W	77.2	303	143	0.47	550	628	570	594
107	100 × 96 × 5 × 8	300	2.05e5	1,994	0	P-P	W	77.2	303	110	0.36	600	680	608	635
108	100 × 96 × 5 × 8	300	2.05e5	1,994	0	P-P	W	77.2	303	61	0.20	680	758	685	724
109	100 × 96 × 5 × 8	300	2.05e5	1,994	0	P-P	W	77.2	303	57	0.19	750	763	691	734
110	100 × 96 × 5 × 8	300	2.05e5	1,994	0	P-P	W	77.2	303	360	1.19	235	396	21	–
111	100 × 96 × 5 × 8	300	2.05e5	1,994	0	P-P	W	77.2	303	320	1.06	440	474	21	–
112	100 × 96 × 5 × 8	300	2.05e5	1,994	0	P-P	W	77.2	303	250	0.82	450	533	405	488
113	100 × 96 × 5 × 8	300	2.05e5	1,994	0	P-P	W	77.2	303	200	0.66	480	576	508	537
114	100 × 96 × 5 × 8	300	2.05e5	1,994	0	P-P	W	77.2	303	150	0.49	552	618	563	586
115	100 × 96 × 5 × 8	300	2.05e5	1,994	0	P-P	W	77.2	303	100	0.33	618	758	623	649
116	100 × 96 × 5 × 8	300	2.05e5	1,994	0	P-P	W	77.2	303	48	0.16	701	816	716	760
117	203 × 203 × 7 × 11	349	2.05e5	1,500	0	F-F	W	29.0	1,626	550	0.34	688	693	646	646
118	100 × 96 × 5 × 8	290	2.05e5	513	5	P-P	W	19.9	506	362	0.71	532	496	517	522
119	100 × 96 × 5 × 8	298	2.05e5	513	5	P-P	W	19.9	520	110	0.21	694	723	707	716
120	100 × 96 × 5 × 8	289	2.05e5	513	5	P-P	W	19.9	504	40	0.08	863	883	861	864
121	100 × 96 × 5 × 8	290	2.05e5	1,272	5	P-P	W	49.2	399	293	0.73	390	500	498	516
122	100 × 96 × 5 × 8	298	2.05e5	1,271	5	P-P	W	49.2	409	251	0.61	474	552	538	550

(continued)

Table 5.1 (continued)

Label	$b_s \times h_s \times t_w \times t_f$ (mm×mm×mm×mm)	f_{y20} (Mpa)	e_{20} (Mpa)	l (mm)	e (mm)	Ends	Axis	λ_{20}	N_{b20} (kN)	P_0 (kN)	μ_0	T_b Meas	T_b FEM	T_b Anal	T_b EC3
124	100 × 96 × 5 × 8	281	2.05e5	2,028	5	P-P	W	78.5	285	174	0.61	509	546	525	551
125	100 × 96 × 5 × 8	281	2.05e5	2,020	5	P-P	W	78.2	286	171	0.60	485	552	530	555
126	100 × 96 × 5 × 8	281	2.05e5	2,023	5	P-P	W	78.3	285	173	0.61	495	548	527	553
127	100 × 96 × 5 × 8	285	2.05e5	2,770	5	P-P	W	107.2	204	127	0.62	457	523	511	548
128	100 × 96 × 5 × 8	290	2.05e5	2,772	5	P-P	W	107.3	206	73	0.35	587	646	600	639
130	100 × 96 × 5 × 8	289	2.05e5	2,772	5	P-P	W	107.3	205	6	0.03	858	1,070	1,055	1,004
131	100 × 96 × 5 × 8	289	2.05e5	3,510	5	P-P	W	135.9	147	105	0.71	446	465	437	522
132	100 × 96 × 5 × 8	296	2.05e5	3,510	5	P-P	W	135.9	149	90	0.61	493	516	510	552
133	100 × 96 × 5 × 8	289	2.05e5	3,510	5	P-P	W	135.9	147	12	0.08	727	899	858	867
BL1	101.85 × 98.85 × 5.92 × 7.61	286.5	2.05e5	513	5	P-P	W	20.0	511	362	0.71	532	511	518	523
CL1	101.78 × 99.07 × 6.43 × 7.8	292.5	2.05e5	513	5	P-P	W	20.2	541	110	0.20	694	735	715	722
DL1	102.28 × 99.12 × 6.13 × 7.68	282.5	2.05e5	513	5	P-P	W	20.0	513	40	0.08	863	882	864	867
BL3	101.93 × 98.9 × 5.97 × 7.64	286.5	2.05e5	1,272	5	P-P	W	49.6	404	292	0.72	390	504	503	519
CL3	101.9 × 99.25 × 6.13 × 7.82	292.5	2.05e5	1,271	5	P-P	W	49.6	422	251	0.59	474	559	544	556
DL3	102.15 × 99.15 × 6.02 × 7.73	282.5	2.05e5	1,269	5	P-P	W	49.4	405	24	0.06	749	959	912	908
SL40	102.15 × 99.15 × 6.02 × 7.73	280	2.05e5	2,020	5	P-P	W	78.6	296	170	0.58	525	550	537	562
SL41	101.84 × 98.97 × 5.73 × 7.58	280	2.05e5	2,026	5	P-P	W	78.9	286	174	0.61	509	538	526	552
SL42	101.82 × 99.04 × 5.76 × 7.61	280	2.05e5	2,020	5	P-P	W	78.7	288	171	0.59	485	544	531	556
SL44	101.68 × 99.17 × 5.73 × 7.6	280	2.05e5	2,023	5	P-P	W	78.9	287	173	0.60	495	540	527	553
AL5	101.94 × 99.06 × 5.78 × 7.68	280	2.05e5	2,770	5	P-P	W	107.6	206	127	0.62	457	521	513	550

(continued)

Table 5.1 (continued)

Label	$b_s \times h_s \times t_w \times t_f$ (mm×mm×mm×mm)	f_{y20} (Mpa)	e_{20} (Mpa)	l (mm)	e (mm)	Ends	Axis	λ_{20}	N_{b20} (kN)	P_0 (kN)	μ_0	T_b Meas	T_b FEM	T_b Anal	T_b EC3
BL5	101.76 × 98.95 × 5.76 × 7.62	286.5	2.05e5	2,772	5	P-P	W	108.0	206	73	0.35	587	622	600	639
CL5	102.03 × 99.25 × 5.98 × 7.76	292.5	2.05e5	2,771	5	P-P	W	107.9	214	34	0.16	587	775	697	760
BL6	101.88 × 98.93 × 5.93 × 7.63	286.5	2.05e5	3,510	5	P-P	W	137.0	149	105	0.71	446	472	443	524
CL6	102.05 × 99.12 × 5.94 × 7.71	292.5	2.05e5	3,510	5	P-P	W	136.6	152	90	0.59	493	516	515	557
DL6	101.68 × 99.17 × 5.73 × 7.6	282.5	2.05e5	3,510	5	P-P	W	136.9	147	12	0.08	727	896	857	866
Lie1	257 × 260 × 11.1 × 17.5	300	2e5	3,810	0	F-F	S	34.2	2,700	1,760	0.65	565	584	562	539
Lie2	254 × 254 × 7.9 × 14.3	300	2e5	3,810	0	F-F	S	34.2	2,124	1,424	0.67	586	579	557	534
Lie3	254 × 254 × 7.9 × 14.3	300	2e5	3,810	0	F-F	S	34.2	2,124	1,424	0.67	584	579	557	534
Ali1	152 × 152 × 6.8 × 8.9	320	2.05e5	1,800	0	P-P	W	47.4	803	186	0.23	701	769	684	703
Ali2	152 × 152 × 6.8 × 8.9	320	2.05e5	1,800	0	P-P	W	47.4	803	373	0.46	626	655	586	597
Ali3	152 × 152 × 6.8 × 8.9	320	2.05e5	1,800	0	P-P	W	47.4	803	559	0.70	557	572	511	527
Ali4	101.2 × 177.8 × 4.8 × 7.9	320	2.05e5	1,800	0	P-P	W	75.1	387	101	0.26	644	746	656	685
Ali5	101.2 × 177.8 × 4.8 × 7.9	320	2.05e5	1,800	0	P-P	W	75.1	387	202	0.52	629	607	554	578
Ali6	101.2 × 177.8 × 4.8 × 7.9	320	2.05e5	1,800	0	P-P	W	75.1	387	303	0.78	539	530	434	502
Ali7	76 × 127 × 4 × 7.6	320	2.05e5	1,800	0	P-P	W	96.7	198	50	0.25	717	731	651	689
Ali8	76 × 127 × 4 × 7.6	320	2.05e5	1,800	0	P-P	W	96.8	198	101	0.51	658	590	550	582
Ali9	76 × 127 × 4 × 7.6	320	2.05e5	1,800	0	P-P	W	96.8	198	151	0.76	567	507	407	507
RS45-1	155.8 × 161.5 × 7.6 × 10.9	326	2.01e5	1,500	1.74	P-P	W	38.2	1,093	709	0.65	647	568	531	540
RS55-1	135.4 × 203.3 × 5.41 × 7.7	357	2.05e5	1,500	3.19	P-P	W	46.8	765	444	0.58	571	573	548	560
RS81-1	88.9 × 152.4 × 4.78 × 8.1	312	2e5	1,500	2.38	P-P	W	70.4	350	261	0.74	499	507	466	513
RS97-1	76.2 × 127.9 × 4.54 × 7.09	320	2e5	1,500	4.08	P-P	W	82.8	233	134	0.58	606	533	532	562

(continued)

Table 5.1 (continued)

Label	$b_s \times h_s \times t_w \times t_f$ (mm×mm×mm×mm)	f_{y20} (Mpa)	e_{20} (Mpa)	l (mm)	e (mm)	Ends	Axis	λ_{20}	N_{b20} (kN)	P_0 (kN)	μ_0	T_b Meas	FEM	Anal	EC3
SCRI1	203.6 × 203.2 × 7.2 × 11	301	2.05e5	2,968	0	R^c-R	W	57.4	1,100	357.5	0.32	646	711	637	652
SCRI2	203.6 × 203.2 × 7.2 × 11	301	2.05e5	2,875	0	R-R	W	55.6	1,121	360.3	0.32	681	712	640	653
SCRI3	203.6 × 203.2 × 7.2 × 11	301	2.05e5	2,875	0	R-R	W	55.6	1,121	360	0.32	668	712	640	653
SCRI4	203.6 × 203.2 × 7.2 × 11	301	2.05e5	3,061	0	R-R	W	59.2	1,080	600	0.56	604	614	551	567
SCRI5	203.6 × 203.2 × 7.2 × 11	301	2.05e5	2,463	0	R-R	W	47.6	1,212	600	0.49	597	636	576	586
SCRI6	203.6 × 203.2 × 7.2 × 11	301	2.05e5	2,745	0	R-R	W	53.1	1,150	600	0.52	613	626	565	578
SCRI7	203.6 × 203.2 × 7.2 × 11	301	2.05e5	3,001	0	R-R	W	58.0	1,093	840	0.77	556	560	466	506
SCRI8	203.6 × 203.2 × 7.2 × 11	301	2.05e5	2,894	0	R-R	W	55.9	1,117	840	0.75	553	564	478	511
SCRI9	203.6 × 203.2 × 7.2 × 11	301	2.05e5	3,001	0	R-R	W	58.0	1,093	840	0.77	530	560	466	506
SCRII1	203.6 × 203.2 × 7.2 × 11	301	2.05e5	2820	0	R-R	W	54.5	1,133	358	0.32	680	715	643	656
SCRII2	203.6 × 203.2 × 7.2 × 11	301	2.05e5	2,849	0	R-R	W	55.1	1,127	360	0.32	670	713	641	654
SCRII3	203.6 × 203.2 × 7.2 × 11	301	2.05e5	2,876	0	R-R	W	55.6	1,121	360	0.32	675	715	643	653
SCRII4	203.6 × 203.2 × 7.2 × 11	301	2.05e5	3,079	0	R-R	W	59.5	1,075	600	0.56	609	614	550	567
SCRII5	203.6 × 203.2 × 7.2 × 11	301	2.05e5	2,883	0	R-R	W	55.7	1,119	600	0.54	604	621	559	573
SCRII6	203.6 × 203.2 × 7.2 × 11	301	2.05e5	2,823	0	R-R	W	54.6	1,132	600	0.53	629	623	561	575
SCRII7	203.6 × 203.2 × 7.2 × 11	301	2.05e5	2,916	0	R-R	W	56.4	1,112	840	0.76	542	563	476	510
SCRII8	203.6 × 203.2 × 7.2 × 11	301	2.05e5	2,916	0	R-R	W	56.4	1,112	840	0.76	550	563	476	510
SCRII9	203.6 × 203.2 × 7.2 × 11	301	2.05e5	2,842	0	R-R	W	54.9	1,128	840	0.74	549	566	484	513

(continued)

Table 5.1 (continued)

Label	$b_s \times h_s \times t_w \times t_f$ (mm×mm×mm×mm)	f_{y20} (Mpa)	e_{20} (Mpa)	l (mm)	e (mm)	Ends	Axis	λ_{20}	N_{b20} (kN)	P_0 (kN)	μ_0	T_b Meas	FEM	Anal	EC3
Choe7	203.7 × 206.2 × 7.9 × 12.6	413	2.05e5	3,500	0	P-P	W	67.3	1,414	1,413	1.00	300	477	102	351
Choe8	203.7 × 206.2 × 7.9 × 12.6	413	2.05e5	3,500	0	P-P	W	67.3	1,414	1,134	0.80	500	523	418	495
Choe9	203.7 × 206.2 × 7.9 × 12.6	413	2.05e5	3,500	0	P-P	W	67.3	1,414	800	0.57	600	588	538	564
Choe10	204.7 × 353.1 × 9.4 × 16.8	406	2.05e5	3,450	0	P-P	W	70.0	2,027	1,435	0.71	500	544	480	524
Choe11	204.7 × 353.1 × 9.4 × 16.8	406	2.05e5	3,450	0	P-P	W	70.0	2,027	1,070	0.53	600	588	550	576

[a]F-Fixed. [b]P-Pinned. [c]R-Rotationally restrained.

Wang and Davies [21] reported 18 tests on rotationally restrained steel column assemblies under fire conditions. The geometrical length of the test column is 3,710 mm. The furnace temperature-time relationship was regulated to follow a linear history, to reach 1,000 °C at 60 min. The test labels in [21] are used in Table 5.1, which are "SCRI1" to "SCRII9". The effective lengths of the test columns, given in Table 5.1, were obtained by an approximated method [21].

Choe et al. [22] conducted tests to determine the fundamental behavior of steel members under fire loading. A total of 11 full-scale steel members were tested under combined thermal and structural loading. Radiant heating and control equipment were used to apply the thermal loading. The 5 column high temperature tests in [22] are considered, which are labeled as "Choe7" to "Choe11" in Table 5.1.

5.4.2 Buckling Temperature of Restrained Column

Simms et al. [4] reported test on structural performance of axially restrained steel columns subjected to elevated temperatures. Table 5.2 gives the parameters for those tests. The test labels in [4] are used in Table 5.2.

Tests on steel columns with finite degree of axial restraint in [5] are given in Table 5.2, which are labeled as "23-0.1-0.2" to "13-0.2-0.6".

Tests on steel columns with axial restraint in [7] are considered, as given in Table 5.2.

5.4.3 Limit Temperature of Restrained Column

Rodrigues et al. [6] conducted tests to study the limit temperature of compressed steel elements with restrained thermal elongation. A total of 168 tests on hinged bars were performed. 36 tests in [6] are considered. The parameters in those tests are presented in Table 5.3. The test labels in [6] are used in Table 5.3.

5.5 Compare Among Predicted and Measured Results

5.5.1 Buckling Temperature for Free Column

The measured and predicted results for buckling temperature of free steel column are also given in Table. 5.1. Results predicted by Eq. 5.18 (marked by "Anal."), Eq. 5.20 (marked by "EC3") and finite element simulation (marked by "FEM") are provided. The measured results are marked by "Meas.".

Figure 5.5 shows the comparisons among predicted and measured results with respect to utilization factor. For most tests, the 3 different calculation methods give

Table 5.2 Measured and predicted buckling temperature for restrained columns

No.	$b_s \times h_s \times t_w \times t_f$ (mm×mm×mm×mm)	f_{y20} (MPa)	e_{20} (MPa)	l (mm)	e (mm)	λ_{20}	P_0 (kN)	k_a (kN/mm)	T_b Meas	T_b FEM	T_b Anal
Simms et al. [4]											
TP112	46 × 80 × 3.8 × 5.2	320	2.05e5	1,600	0	133.8	43	4.1	233	204	107(190)[a]
TP102	46 × 80 × 3.8 × 5.2	320	2.05e5	1,600	0	133.8	0	4.1	536	473	426(501)
Ali et al. [5]											
23-0.1-0.2	152 × 152 × 6.8 × 8.9	320	2.05e5	1,800	0	47.4	186	41.2	640	560	500
23-0.1-0.4	152 × 152 × 6.8 × 8.9	320	2.05e5	1,800	0	47.4	373	41.2	598	511	427
23-0.1-0.6	152 × 152 × 6.8 × 8.9	320	2.05e5	1,800	0	47.4	559	41.2	547	461	304
23-0.2-0.2	152 × 152 × 6.8 × 8.9	320	2.05e5	1,800	0	47.4	186	82.4	583	474	384
23-0.2-0.4	152 × 152 × 6.8 × 8.9	320	2.05e5	1,800	0	47.4	373	82.4	517	436	298
23-0.2-0.6	152 × 152 × 6.8 × 8.9	320	2.05e5	1,800	0	47.4	559	82.4	363	369	212
19-0.1-0	101.2 × 177.8 × 4.8 × 7.9	320	2.05e5	1,800	0	75.1	0	27.1	552	542	486
19-0.1-0.2	101.2 × 177.8 × 4.8 × 7.9	320	2.05e5	1,800	0	75.1	101	27.1	555	491	406
19-0.1-0.4	101.2 × 177.8 × 4.8 × 7.9	320	2.05e5	1,800	0	75.1	202	27.1	466	440	297
19-0.1-0.6	101.2 × 177.8 × 4.8 × 7.9	320	2.05e5	1,800	0	75.1	303	27.1	364	365	181
19-0.2-0	101.2 × 177.8 × 4.8 × 7.9	320	2.05e5	1,800	0	75.1	0	54.1	507	535	333
19-0.2-0.2	101.2 × 177.8 × 4.8 × 7.9	320	2.05e5	1,800	0	75.1	101	54.1	455	393	268
19-0.2-0.4	101.2 × 177.8 × 4.8 × 7.9	320	2.05e5	1,800	0	75.1	202	54.1	432	333	202
19-0.2-0.6	101.2 × 177.8 × 4.8 × 7.9	320	2.05e5	1,800	0	75.1	303	54.1	408	265	136
13-0.1-0	76 × 127 × 4 × 7.6	320	2.05e5	1,800	0	96.6	0	18.2	445	472	406

(continued)

Table 5.2 (continued)

No.	$b_s \times h_s \times t_w \times t_f$ (mm×mm×mm×mm)	f_{y20} (Mpa)	e_{20} (Mpa)	l (mm)	e (mm)	λ_{20}	P_0 (kN)	k_a (kN/mm)	T_b Meas	T_b FEM	T_b Anal
13-0.1-0.2	76 × 127 × 4 × 7.6	320	2.05e5	1,800	0	96.6	51	18.2	536	425	324
13-0.1-0.4	76 × 127 × 4 × 7.6	320	2.05e5	1,800	0	96.6	101	18.2	333	354	239
13-0.1-0.6	76 × 127 × 4 × 7.6	320	2.05e5	1,800	0	96.6	151	18.2	386	281	153
13-0.2-0	76 × 127 × 4 × 7.6	320	2.05e5	1,800	0	96.6	0	36.5	530	339	260
13-0.2-0.2	76 × 127 × 4 × 7.6	320	2.05e5	1,800	0	96.6	51	36.5	441	293	211
13-0.2-0.4	76 × 127 × 4 × 7.6	320	2.05e5	1,800	0	96.6	101	36.5	410	248	162
13-0.2-0.6	76 × 127 × 4 × 7.6	320	2.05e5	1,800	0	96.6	151	36.5	336	202	114
Tan et al. [7]											
RS45-2	155.7 × 161.5 × 7.6 × 11	326	2.01e5	1,500	2.94	44.3	709	33.4	645	495	447
RS45-3	155.7 × 162 × 7.6 × 11	326	2.01e5	1,500	6.49	44.3	703	41.0	499	463	433
RS55-2	135.5 × 203.1 × 5.41 × 7.6	357	2.05e5	1,500	3.43	54.3	440	22.1	519	501	467
RS55-3	135.5 × 203.2 × 5.41 × 7.7	357	2.05e5	1,500	5.05	54.2	440	31.4	459	470	437
RS55-4	133.7 × 204.4 × 5.41 × 7.46	344	2.21e5	1,500	3.67	55.4	387	39.7	417	465	419
RS81-2	90.1 × 154 × 5.16 × 7	295	2.15e5	1,500	0.65	83.8	195	18.3	510	454	322
RS81-3	90.5 × 154.7 × 4.54 × 7.9	332	2.08e5	1,500	1.99	79.9	260	25.4	378	373	254
RS81-4	89.9 × 154.1 × 5.16 × 7.1	305	2.15e5	1,500	6.29	83.8	260	31.4	378	198	171
RS97-2	76.8 × 128.3 × 4.36 × 7.2	320	2e5	1,500	4.16	94.4	134	14.7	399	372	318
RS97-3	76 × 127.8 × 4.54 × 7.1	316	2e5	1,500	1.54	96.3	179	21.6	292	275	168
RS97-4	76.4 × 129.9 × 4.54 × 7.94	316	2e5	1,500	1.65	94.2	136	30.1	330	326	236

[a]The value in bracket is obtained by using measured effective length for calculation.

Table 5.3 Measured and predicted limit temperature for restrained steel components in Rodrigues et al. [6]

No.	$b \times t$ (mm×mm)	f_{y20} (Mpa)	e_{20} (Mpa)	l (mm)	e (mm)	λ_{20}	P_0 (kN)	μ_0	k_a (kN/mm)	β_l	$T_{b,EC3}$ °C	T_{lim} Eq.5.21	$Meas$
166	50 × 8	327.5	2.1e5	460	1	199.2	8.88	0.53	0	0	522	522	578
48	50 × 8	319.7	2.1e5	460	1	199.2	8.88	0.54	1	0.01	522	498	546
86	50 × 8	324.8	2.1e5	460	1	199.2	8.88	0.53	10	0.05	522	393	387
100	50 × 8	319.9	2.1e5	460	1	199.2	8.88	0.54	24	0.13	522	392	376
135	50 × 8	324.8	2.1e5	460	1	199.2	8.88	0.53	42	0.23	522	393	382
158	50 × 8	327.5	2.1e5	460	1	199.2	8.88	0.53	98	0.54	522	393	382
173	50 × 8	327.5	2.1e5	460	8	199.2	4.47	0.27	0	0	568	568	574
61	50 × 8	319.7	2.1e5	460	8	199.2	4.47	0.27	1	0.01	568	552	548
77	50 × 8	324.8	2.1e5	460	8	199.2	4.47	0.27	10	0.05	568	480	529
109	50 × 8	319.9	2.1e5	460	8	199.2	4.47	0.27	24	0.13	568	480	538
126	50 × 8	319.9	2.1e5	460	8	199.2	4.47	0.27	42	0.23	568	480	528
186	50 × 8	327.5	2.1e5	460	8	199.2	4.47	0.27	98	0.54	568	480	523
148	50 × 12	322.7	2.1e5	460	1	132.8	26.21	0.54	0	0	521	521	600
50	50 × 12	314.4	2.1e5	460	1	132.8	26.21	0.54	1	0.00	521	502	539
88	50 × 12	318.6	2.1e5	460	1	132.8	26.21	0.54	10	0.04	521	368	492
98	50 × 12	322.7	2.1e5	460	1	132.8	26.21	0.54	24	0.09	521	369	379
140	50 × 12	322.7	2.1e5	460	1	132.8	26.21	0.54	42	0.15	521	369	364
150	50 × 12	317.9	2.1e5	460	1	132.8	26.21	0.54	98	0.36	521	368	385
175	50 × 12	317.9	2.1e5	460	12	132.8	10.47	0.22	0	0	582	582	552

(continued)

Table 5.3 (continued)

No.	$b \times t$ (mm×mm)	f_{y20} (Mpa)	e_{20} (Mpa)	l (mm)	e (mm)	λ_{20}	P_0 (kN)	μ_0	k_a (kN/mm)	β_l	$T_{b,EC3}$ °C	T_{lim} Eq.5.21	Meas
66	50 × 12	314.4	2.1e5	460	12	132.8	10.47	0.22	1	0.00	582	571	545
75	50 × 12	318.6	2.1e5	460	12	132.8	10.47	0.22	10	0.04	582	489	530
113	50 × 12	318.6	2.1e5	460	12	132.8	10.47	0.22	24	0.09	582	489	528
122	50 × 12	322.7	2.1e5	460	12	132.8	10.47	0.22	42	0.15	583	489	527
183	50 × 12	317.9	2.1e5	460	12	132.8	10.47	0.22	98	0.36	582	489	524
145	50 × 20	298.9	2.1e5	460	1	79.7	84.37	0.57	0	0	517	517	582
53	50 × 20	313.3	2.1e5	460	1	79.7	84.37	0.55	1	0.00	520	507	569
93	50 × 20	290.8	2.1e5	460	1	79.7	84.37	0.58	10	0.02	516	387	509
96	50 × 20	290.8	2.1e5	460	1	79.7	84.37	0.58	24	0.05	516	339	408
143	50 × 20	298.9	2.1e5	460	1	79.7	84.37	0.57	42	0.09	517	343	357
154	50 × 20	291.4	2.1e5	460	1	79.7	84.37	0.58	98	0.21	516	339	343
177	50 × 20	291.4	2.1e5	460	12	79.7	25.03	0.17	0	0	598	598	549
67	50 × 20	313.3	2.1e5	460	12	79.7	25.03	0.16	1	0.00	601	594	570
71	50 × 20	292.7	2.1e5	460	12	79.7	25.03	0.17	10	0.02	598	528	538
117	50 × 20	290.8	2.1e5	460	12	79.7	25.03	0.17	24	0.05	597	502	532
119	50 × 20	298.9	2.1e5	460	12	79.7	25.03	0.17	42	0.09	599	504	520
181	50 × 20	291.4	2.1e5	460	12	79.7	25.03	0.17	98	0.21	598	503	522

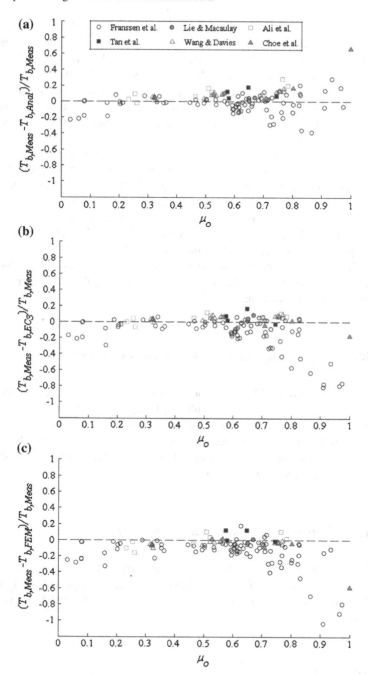

Fig. 5.5 Compare among results for buckling temperature of free steel columns with respect to utilization factor. **a** Anal. versus Meas. **b** EC3 versus Meas. **c** FEM versus Meas

acceptable results (errors are within 20 %). For tests with high utilization factor ($\mu_0 > 0.83$), neither Eq. 5.20 nor FEM gives acceptable predictions. Because the approach for calculating buckling resistance $N_{b,T}$ in Eq. 5.18 is based on fitting with test data reported by Franssen et al. [13], as expected Eq. 5.18 gives comparatively better predictions at high utilization factor (except one test, all tests at this range are reported by Franssen et al. [13]). For most tests with low utilization factor ($\mu_0 < 0.16$), the differences among predicted and measured results can be greater than 100 °C, as shown in Fig. 5.6.

For tests reported by Franssen et al. [13], predicted results are higher than the measured values for most cases. For the remaining tests, Eqs. 5.18 and 5.20 give consistent and good predictions (comparatively Eq. 5.20 gives better predictions, as shown in Fig. 5.6), and the predicted results are lower than the measured values for most cases. FEM also gives good predictions for those remaining tests, and the predicted results are higher than the measured values for most cases. Overall, Eq. 5.20 gives best predictions at middle utilization factors ($0.2 < \mu_0 < 0.8$).

Figure 5.7 shows the comparisons among predicted and measured results with respect to non-dimensional slenderness. The accuracies of calculation methods are not sensitive to non-dimensional slenderness.

5.5.2 Buckling Temperature for Restrained Column

The measured and predicted results for buckling temperature of restrained steel column are also given in Table. 5.2. Results predicted by Eq. 5.18 and finite element simulation are provided. Figure 5.8 shows the comparisons among predicted and measured results. Except 4 tests reported by Ali et al. [5], Eq. 5.18 fails to give acceptable prediction for the other 31 tests (errors are larger than 20 %). For most tests, FEM gives acceptable predictions.

5.5.3 Limit Temperature for Restrained Column

The measured and predicted results for limit temperature of restrained steel column are also given in Table. 5.3. Results predicted by Eq. 5.21 are provided. Figure 5.9 shows the comparisons among predicted and measured results, which shows good agreement.

5.5.4 Discussion

When deriving formula for calculating buckling resistance of axially loaded steel column in fire, the severity factor in Eq. 5.16 is taken as 0.65, which leads to 50 %

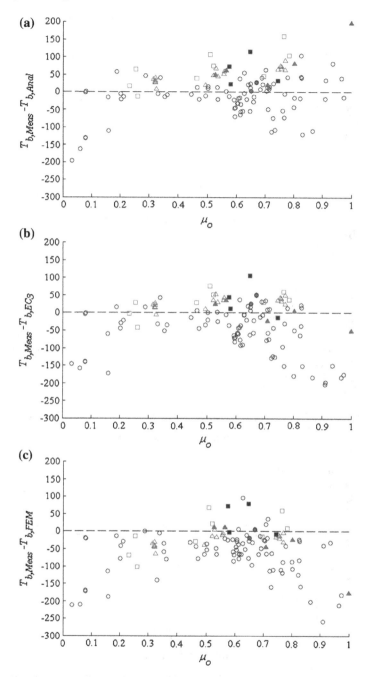

Fig. 5.6 Differences among predicted and measured buckling temperatures for free steel columns. **a** Anal. versus Meas. **b** EC3 versus Meas. **c** FEM versus Meas

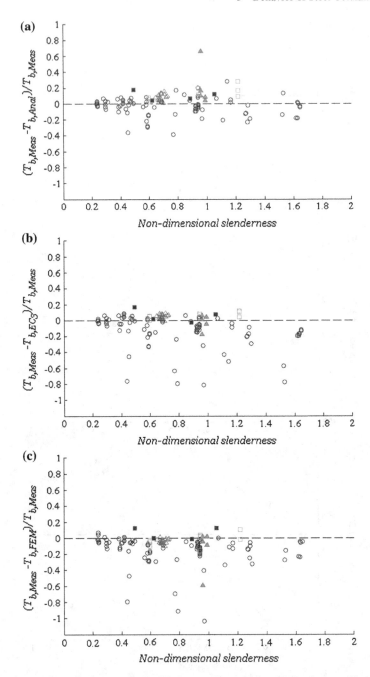

Fig. 5.7 Compare among results for buckling temperature of free steel columns with respect to non-dimensional slenderness. **a** Anal. versus Meas. **b** EC3 versus Meas. **c** FEM versus Meas

Fig. 5.8 Compare among results for buckling temperature of restrained steel columns

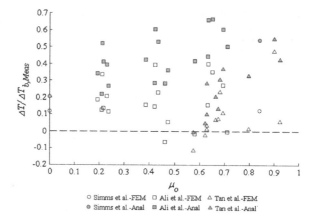

Fig. 5.9 Compare among results for limit temperature of restrained steel columns

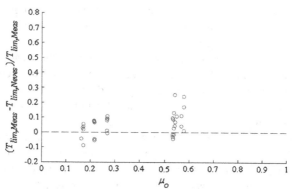

of safe results [13]. In other words, for tests summarized by by Franssen et al. [13], buckling resistance predicted by Eq. 5.13 might be greater or less than the measured value, and the probability of over- and under-prediction by Eq. 5.13 are equal as 50 %. As a result, Eqs. 5.18 and 5.20, which are based on Eq. 5.13, might give unacceptable prediction of buckling temperatures for free steel columns.

In Eq. 5.13, the effects of initial geometric imperfection, load eccentricity and residual stresses are considered. However, in the FEM model, the effect of residual stresses is not considered and the effects of initial geometric imperfection and load eccentricity are simply considered. Also, the temperatures within test columns are not uniformly distributed but in FEM model uniform temperature distributions are assumed. Besides, for tests summarized by [13], in FEM model pinned-pinned end conditions and effective column lengths are used for all tests. These treatments and assumptions, along with errors in material models, result in divergence among the FEM results and measured data for buckling temperatures of free columns.

When using Eq. 5.18 to calculate buckling temperatures of axially restrained steel columns, buckling lengths are taken as equal to the geometrical lengths of the test columns. However, as reported in Simms et al. [4], the buckling length of axially

Fig. 5.10 Measured and predicted axial forces for tests reported by Simms et al. [4]

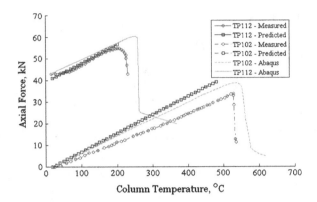

restrained column is not the same as the column length (In their tests, the effective length is $0.893l_0$, where l_0 is column length). The results for buckling temperature predicted by Eq. 5.18 with using the reported effective length are also provided in Table 5.2, which shows good agreement with the measured values. Taking effective lengths of axially restrained steel columns as equal to column lengths might be one significant reason causing the great divergence among the predicted buckling temperatures by Eq. 5.18 and measured data for restrained columns. Figure 5.10 shows the comparison among results for axial forces predicted by Eq. 5.19 and measured values for tests reported by Simms et al. [4]. The numerical results obtained from Abaqus by Wang [23] are also provided. Equation 5.19 fits well with test TP112, but diverges considerably with test TP102. However, Eq. 5.19 agree with numerical results for both tests. The error in predicting axial force or additional restraining force also leads to divergence among the predicted and measured buckling temperatures for restrained steel columns.

5.6 Professional Factor for Limit Temperature

Professional factor is used to account for model error in predicting limit temperature of steel columns, which is defined as the ratio of measured and predicted results.

The professional factor for Eq. 5.20 is characterized by using free column test data in Table 5.1, which has a mean of 0.949 and a COV of 0.016, and can be best described by the extreme value distribution as shown in Fig. 5.11. The probability density function (pdf) for the extreme value distribution is given by

$$f(x) = \sigma^{-1}exp(\frac{x - \mu}{\sigma})exp(-exp(\frac{x - \mu}{\sigma})) \tag{5.22}$$

where $\mu = 1.00378$ is location parameter and $\sigma = 0.0954518$ is scale parameter.

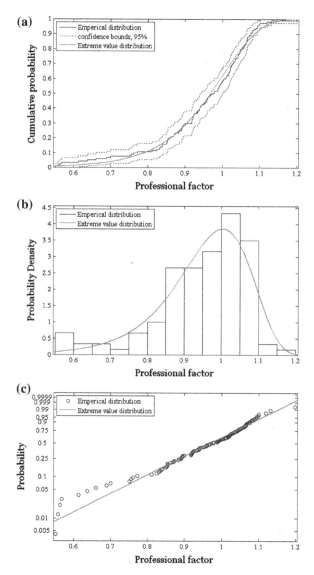

Fig. 5.11 Probabilistic property for professional factor of Eq. 5.20. **a** Cumulative probability. **b** Probability density. **c** Probability plot

The professional factor for Eq. 5.18 is characterized by using free column test data in Table 5.1, which has a mean of 1.018 and a COV of 0.013, and can be well described by the normal, gamma or lognormal distribution as shown in Fig. 5.12.

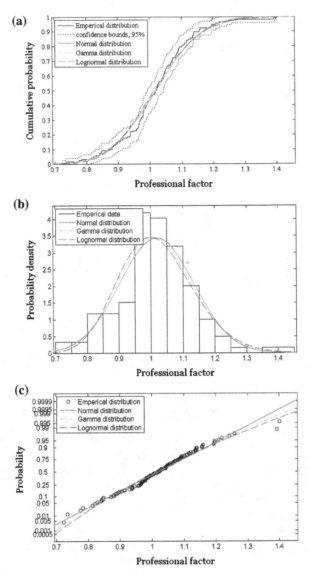

Fig. 5.12 Probabilistic property for professional factor of Eq. 5.18. **a** Cumulative probability. **b** Probability density. **c** Probability plot

The professional factor for Eq. 5.21 is characterized by using restrained column test data in Table 5.3, which has a mean of 0.949 and a COV of 0.006, and can be best described by the normal distribution as shown in Fig. 5.13.

Fig. 5.13 Probabilistic property for professional factor of Eq. 5.21

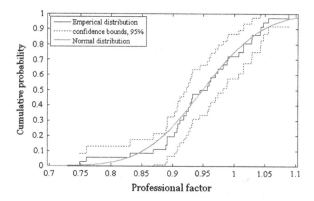

5.7 Conclusions

Comparing with test data reported in literature, the accuracy of simple approaches for predicting the buckling and limiting temperatures of steel columns in fire conditions have been investigated. Based on the results of this study, the following conclusions can be drawn:

- Overall, the calculation approaches adopted in eurocode, and the FEM model give acceptable prediction of buckling temperatures for fire tests on free steel columns reported in literature (errors are within 20 %). For tests with high utilization factor($\mu_0 > 0.83$), neither the simple closed-form formula given in eurocode Eq. 5.20 nor the FEM model gives acceptable results. For tests with low utilization factor ($\mu_0 < 0.16$), all 3 approaches fail to give good prediction for most cases (the over-predictions are higher than 100 °C). The professional factor for the simple closed-form formula in eurocode Eq. 5.20 has a mean of 0.949 and a COV of 0.016 and can be best described by a extreme value distribution. The professional factor for the analytical approach Eq. 5.18 has a mean of 1.018 and a COV of 0.013, and can be well described by either a normal, gamma or lognormal distribution.
- The simple approach fails to give acceptable prediction of buckling temperatures for axially restrained steel columns in fire conditions. The investigated FEM model gives acceptable prediction of buckling temperature for axially restrained steel columns.
- The approach proposed by Neves et al. [14] Eq. 5.21 gives acceptable prediction of limiting temperature of axially restrained steel components. The professional factor of the approach has a mean of 0.949 and a COV of 0.006, and can be best described by the normal distribution.

References

1. B. 476–20, *Fire Tests on Building Materials and Structures, Part 20: Methods for Determination of the Fire Resistance of Elements of Construction (General Principles)* (British Standards Institution, London, 1987)
2. J. Quintiere, *Fundamentals of Fire Phenomena* (Wiley, USA, 2006)
3. B. Kirby, *The Behaviour of a Multi-Story Steel Framed Building Subjected to Fire Attack, Experimental Data* (British Steel, UK, 1998)
4. W. Simms, D. O'Connor, F. Ali., M. Randall. J. Appl. Fire Sci. **5**, 269 (1996)
5. F. Ali, P. Shepherd, M. Randall, I. Simms, D. O'Connor, I. Burgess, J. Constr. Steel Res. **46**, 305 (1998)
6. J.C. Rodrigues, I.C. Neves, J. Valente, Fire Saf. J. **35**, 77 (2000)
7. K. Tan, W. Toh, Z. Huang, G. Phng, Eng. Struct. **29**, 1641 (2007)
8. G. Li, P. Wang, Y. Wang, J. Constr. Steel Res. **66**, 1138 (2010)
9. I.C. Neves, Fire Saf. J. **24**, 211 (1995)
10. Y. Wang, J. Struct. Eng. **130**, 371 (2004)
11. J. Franssen, Fire Saf. J. **34**, 191 (2000)
12. G. Li, C. Zhang, J. Constr. Steel Res. (2011)
13. J. Franssen, J. Schleich, L. Cajot, W. Azpiazu, J. Constr. Steel Res. **37**, 175 (1996)
14. I.C. Neves, J. Valente, J.C. Rodrigues, Fire Saf. J. **37**, 753 (2002)
15. P. Wang, G. Li, Y. Wang, J. Constr. Steel Res. **66**, 1422 (2010)
16. Y. Anderberg, in *Proceedings of the Fifth International Conference on Structures in Fire*, Singapore, 2008, pp. 253–265
17. Y. Wang, T. Lennon, D. Moore, J. Constr. Steel Res. **35**, 291 (1995)
18. J. Franssen, J. Schleich, L. Cajot, J. Constr. Steel Res. **35**, 49 (1995)
19. BSI, *Eurocode 3: Design of Steel Structures–Part 1–2: General Rules–Structural Fire Design* (British Standard, London, 2005)
20. T. Lie, B. Macaulay, Evaluation of the fire resistance of protected steel columns. Technical report (1989)
21. Y. Wang, J. Davies, J. Constr. Steel Res. **59**, 291 (2003)
22. L. Choe, A. Varma, A. Agarwal, A. Syrovek, J. Struct. Eng. **137**, 954 (2011)
23. P. Wang, Experimental and theoretical study on behaviors of restrained steel columns in fire. Ph.D. thesis, Tongji University, 2008

Chapter 6
Reliability Analysis

6.1 Introduction

This chapter presents the probabilistic study on reliability of steel columns protected by intumescent coatings to natural fires. Approaches given in Chaps. 2–5 are used as deterministic models. Monte Carlo simulation is adopted for probabilistic analysis.

6.2 Deterministic Approaches

The concept of equivalent fire severity, introduced in Chap. 2, is used to model natural fires. The formula given in eurocode or Eq. 2.40 is used to calculate the equivalent standard fire duration of natural fires.

The simple formula proposed in Chap. 3, Eq. 3.27 is adopted to calculate the maximum temperature of insulated steel columns in natural fires.

The thermal insulation property of intumescent coating is assessed by the concept of equivalent constant therm resistance proposed in Chap. 4. The calculated constant thermal resistance of intumescent coatings given in Table 4.1 are adopted to consider aging effect.

The simple formula given in Eurocode or Eq. 5.20 in Chap. 5 is used to calculate the buckling temperature of steel columns in fire conditions.

6.3 Probabilistic Approach

6.3.1 Parameter Uncertainties

Characterization of parameter uncertainties is of fundamental importance in a probabilistic approach. In modeling a post-flashover fire, the major source of

© Springer-Verlag Berlin Heidelberg 2015

C. Zhang, *Reliability of Steel Columns Protected by Intumescent Coatings Subjected to Natural Fires*, Springer Theses, DOI 10.1007/978-3-662-46379-6_6

Table 6.1 Statistics of uncertain parameters

Parameters	Unit	PDF	Mean	COV
q_f	MJ/m^2	Lognormal	100–600	0.3
W	m	Deterministic	3.6/4.8	–
D	m	Deterministic	4.8/6.0	–
H	m	Deterministic	2.7/3.6	–
b	Ws$^{1/2}$/m^{2}K	Normal	2,014	0.09
W_o	m	Deterministic	3/4.5	–
H_o	m	Deterministic	1.5/1.8	–
ζ	–	Lognormal	0.2	1
d_i/k_i	Km^2/W	Lognormal	R_{con}	0.3
B_s	mm	Normal	nominal	0.05
H_s	mm	Normal	nominal	0.05
t_w	mm	Normal	nominal	0.05
t_f	mm	Normal	nominal	0.05
l_c	m	Normal	2.5/3.5	0.05
f_{y20}	MPa	Normal	235	0.063
E_{20}	MPa	Normal	2.05e5	0.045
P_T	N	Normal	$\mu_0{*}N_{b,0}$	0.3

parameter uncertainties is associated with the prescription of fire load, vent, and thermal inertia of the compartment boundary. Based on fire load surveys in different countries, several probabilistic models (e.g. Lognormal distribution [1, 2] and Gumbel type I distribution [3]) had been derived to characterize the distribution of fire load densities. In probabilistic codes, mean and COV of fire load density in different occupancies are given. For example, in Eurocode [4], the fire load density in office building has a mean of 420 MJ/m^2 and a COV of 0.3, and in JCSS probabilistic model code [5], the fire load density in office building has a mean of 600 MJ/m^2 and a COV of 0.3.

When calculating opening factor in deterministic approach, it is assumed that ordinary window glass is immediately destroyed when fire breaks out. However, in many conditions it is possible that some of the openings are partially or fully closed in fire conditions. In JCSS probabilistic model [5], a random parameter ζ is defined to consider the uncertainty of openings in fire conditions. The parameter is described by a truncated lognormal distribution with a mean of 0.2 and a COV of 1. Correspondingly, the effective opening factor in fire conditions is given by

$$F_{o,eff} = F_o(1 - \zeta) \tag{6.1}$$

In compartment fire modeling [6], when calculating heat transfer from fire environment to compartment boundary, the semi-infinite solid assumption is adopted, or in other words the fire temperature is assumed to be not affected by thethickness of the compartment boundary. Therefore, only the uncertainty in evaluating thermal inertia of the compartment boundary is considered. In the following studies, compartment boundaries are assumed to be constructed from normal weight concrete (NWC) only. In [7], the thermal inertia of NWC is described by a normal distribution with a mean of 2,014 Ws$^{1/2}$/m^{2}K and a COV of 0.09.

As been described in Chap. 4, intumescent coatings are reactive materials that they will swell under heating conditions. Due to different chemical components and complex intumescing behavior, the thermal insulation property of different type of intumescent coatings might be different significantly. As a result, although a number of studies on intumescent coatings have been reported in literature, available data to characterized the thermal insulation property of intumescent coatings is limited. In [7], the thermal conductivity of traditional fire protection materials and the thickness of the protection are both described by lognormal distribution. According to the definition of thermal resistance, which is the ratio of protection thickness to thermal conductivity, from probabilistic theory we know if protection thickness and thermal conductivity both follow lognormal distribution, then thermal resistance also follows lognormal distribution. As a result, in our study the constant thermal resistance of intumescent coatings is assumed to follow lognormal distribution with COV of 0.3.

Table 6.1 gives the statistics of uncertain parameters. The distributions of yield strength and elastic modulus of steel at room temperature are assumed to be normal. The COVs for yield strength and elastic modulus used in [8] are adopted. The distributions of the dimensions of the steel columns are also assumed to be normal. The distribution of column service load is assumed to be normal.

6.3.2 Professional Factors

As given in in Sect. 3.5.5, the professional factor for maximum steel temperature calculation by Eq. 3.27, *PF1*, has a mean of 0.955 and a COV of 0.014, and can be best described by lognormal distribution as shown in Fig. 3.8.

As given in Sect. 5.6, the professional factor for buckling steel temperature calculation by Eq. 5.20, *PF2*, has a mean of 0.949 and a COV of 0.016, and can be best described by the extreme value distribution as shown in Fig. 5.11.

6.4 Reliability Theory

Reliability is defined as the ability of a structure or component to perform its required functions under stated conditions for a specified period of time [9]. Take $G = G(X_1, X_2, X_3, \ldots)$ as the performance function, then if $G > 0$ the structure or component can perform its required functions (or the structure or component is safe), and if $G \leq 0$ the structure or component can not (or the structure or component is failure). Correspondingly, the degree of reliability of the structure or component is given by

$$P_s = P[G(X_1, X_2, X_3, \ldots) > 0] = 1 - P[G(X_1, X_2, X_3, \ldots) \leq 0] = 1 - P_{fail}$$
(6.2)

where $X_1, X_2, X_3, \ldots$ are the basic variables; and P_{fail} is the failure probability of the structure or component, given by

$$P_{fail} = P[G(\mathbf{X}) \leq 0] = \int \ldots \int_{G(\mathbf{X}) \leq 0} f_{\mathbf{X}}(\mathbf{x}) d\mathbf{x}$$
(6.3)

where $f_{\mathbf{X}}(\mathbf{x})$ is the joint probability density function. When the basic variables are independent with each other $f_{\mathbf{X}}(\mathbf{x}) = \prod_{i=1}^{n} f_{X_i}(x_i)$, in which $f_{X_i}(x_i)$ is the probability density function (pdf) for the basic variable X_i.

Equation 6.3 shows that the failure probability of a structure or component can be calculated from integration over the failure domain ($G(\mathbf{X}) \leq 0$). The failure probability can also be calculated by simulation methods, thus

$$P_{fail} = \frac{n(G \leq 0)}{N}$$
(6.4)

where N is the total number of simulations; and n is the number of simulations for which $G \leq 0$.

If the pdf of $Z = G(X)$ is normal distribution with the mean of μ_Z and standard deviation of σ_Z, the failure probability can be calculated by using a "reliability index", thus

$$P_{fail} = P[Z \leq 0] = P[\frac{Z - \mu_Z}{\sigma_Z} \leq \frac{0 - \mu_Z}{\sigma_Z}] = \Phi(-\frac{\mu_Z}{\sigma_Z}) = \Phi(-\beta)$$
(6.5)

here

$$\beta = \frac{\mu_Z}{\sigma_Z}$$
(6.6)

is the reliability index. Equation 6.5 shows that the greater β, the smaller is P_{fail}, and vice verse. From Eq. 6.5, we get

$$\beta = \Phi^{-1}(1 - P_{fail})$$
(6.7)

Generally, FORM [9] is usually adopted as approximate method for calculating the reliability index. The procedure is as follows: first, the limit state function is recast in the standard normal space by using a probabilistic transformation if necessary; then find the design point, which minimizes the distance from the origin of the standard normal space to the limit state surface. The reliability index is defined as the obtained distance.

6.5 Reliability Analysis

6.5.1 Performance Function

When exposed to fire, the steel columns will survive if the maximum temperatures they reached are less than their buckling temperatures. The performance function for reliability analysis of steel columns in fire is given by

$$G(\mathbf{X}) = R - S = T_b - T_{s\max} \tag{6.8}$$

6.5.2 Investigated Cases

Table 6.2 gives the investigated cases considered in reliability analysis. Two fire compartments are considered and labeled as "Comp1" and "Comp2", respectively. "Comp1" is 3.6 m width, 4.8 m depth and 3.0 m height, and has a window of 3.0 m width and 1.5 m height. "Comp2" is 4.8 m width, 6.0 m depth and 3.6 m height, and has a window of 4.5 m width and 1.8 m height. The two compartments both have fire doors which are assumed to be closed in fire condition.

In each case, the design floor fire load density, q_f, ranges from 100 to 600 MJ/m^2 with increment of 100 MJ/m^2. The investigated q_f covers the range of type fire compartments given in JCSS model [5].

Three different steel cross sections are considered and labeled as "S1", "S2" and "S3", respectively. "S1" is $H200 \times 200 \times 8 \times 12$, "S2" is $H300 \times 300 \times 10 \times 15$, and "S3" is $H400 \times 400 \times 15 \times 20$. In "Comp1" the column length is 2.5 m and in "Comp2" the column length is 3.5 m. The columns are pinned-pinned, and four sides exposed to fire. The nominal values of steel yield strength and elastic modulus are 235 and 205,000 MPa, respectively. The considered slenderness ratio ranges from 24.8 to 68.9 (along weak axis).

In each case, the utilization factor or load ratio, μ_0, ranges from 0.1 to 0.6. In normal design, load ratios ranging from 0.3 to 0.6 are usually adopted. The probability of coincidence of a fire with maximum values of live load, snow, wind, or earthquake

Table 6.2 Investigated cases in reliability analysis

Label	Comp.	Sec.	λ_{20}	Coat.	Label	Comp.	Sec.	λ_{20}	Coat.
Case 1	Comp1	S1	50.2	C1	Case 7	Comp2	S1	68.9	C1
Case 2	Comp1	S1	50.2	C2	Case 8	Comp2	S1	68.9	C2
Case 3	Comp1	S2	32.9	C1	Case 9	Comp2	S2	46.1	C1
Case 4	Comp1	S2	32.9	C2	Case 10	Comp2	S2	46.1	C2
Case 5	Comp1	S3	24.8	C1	Case 11	Comp2	S3	35.0	C1
Case 6	Comp1	S3	24.8	C2	Case 12	Comp2	S3	35.0	C2

loads is negligible, and a structure is likely to be loaded to only a fraction of the design live load when a fire occurs [10]. Considering the combination of dead and live loads, and assuming the live load in fire condition is half of the design live load, then the lower and upper limit of load ratio in fire condition are 0.15 and 0.6 for normal design with load ratios ranging from 0.3 to 0.6. As a result, the considered load ratios cover the typical values in fire designs.

Two fire protections using intumescent coating are considered and are labeled as "C1" and "C2", respectively. The coating thickness for "C1" is 1 mm and for "C2" is 2 mm. The performance of the intumescent coatings in different design fires are assumed to be the same as that in standard fire, and the insulation properties of the coatings are assumed to be equal to the values reported in [11, 12]. Table 4.1 in Chap. 4 gives the constant thermal resistance of the coatings.

6.5.3 Monte Carlo Simulations

Figure 6.1 shows the flowchart for monte carlo simulations. For each loop, maximum steel temperature calculated by Eq. 3.27 (multiply by profession factor, $PF1$) is compared with failure temperature of steel column calculated by Eq. 5.20 (multiply by profession factor, $PF2$). The number of simulations, N, is taken as 1,000,000. Sensitive study shows that using more simulation loops, e.g. 2,000,000, 5,000,000, yields similar results. Figure 6.2 shows the sampling history of some statistic parameters. The theoretical distributions are also plotted.

6.5.4 Results

Figures 6.3, 6.4, 6.5, 6.6, 6.7, 6.8, 6.9, 6.10, 6.11, 6.12, 6.13, 6.14 show some results of reliability index for the investigated cases. It can been seen that the reliability index, β, decreases with load ratio, μ_0, increases. With service year increases, due to aging effect of intumescent coating, β decreases. As shown in Figs. 6.15 and 6.16 , the aging effect on β is comparatively more serious for cases with small β or high μ_0 than for cases with big β or low μ_0.

For cases with $\beta \geq 1.5$ or $\mu_0 \leq 0.3$, the amount of the decrease of β due to aging effect, $\Delta\beta$, is less than 0.2 (the corresponding increase in failure probability, ΔP_{fail}, is less than 3 %, as shown in Fig. 6.17). For cases with $\beta < 1.5$ or $\mu_0 > 0.3$, the maximum $\Delta\beta$ is about 0.24 (the corresponding maximum ΔP_{fail} is about 9 %).

Fig. 6.1 Flowchart for
Monte Carlo simulations

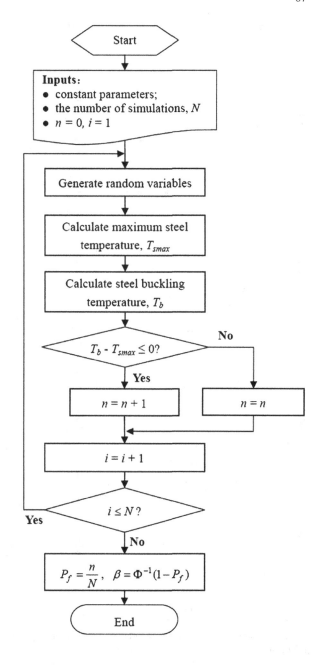

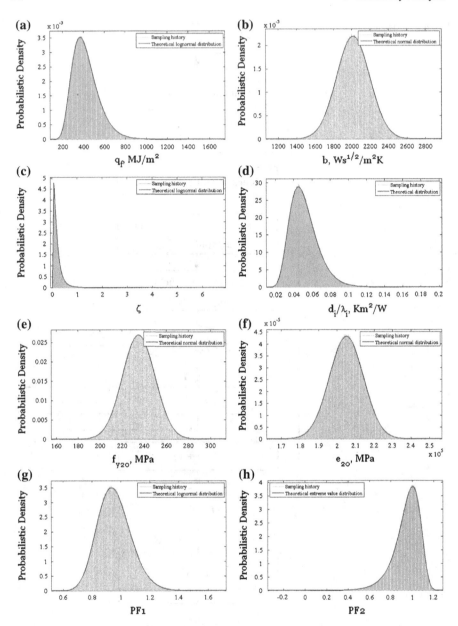

Fig. 6.2 Sampling history versus theoretical distribution. **a** q_f, **b** b, **c** ζ, **d** d_i/λ_i, **e** f_{y20}, **f** e_{20}, **g** PF_1, **h** PF_2

Fig. 6.3 Some results of
reliability index for Case 1.
a $q_f = 200\,\mathrm{MJ/m^2}$,
b $q_f = 300\,\mathrm{MJ/m^2}$,
c $q_f = 400\,\mathrm{MJ/m^2}$

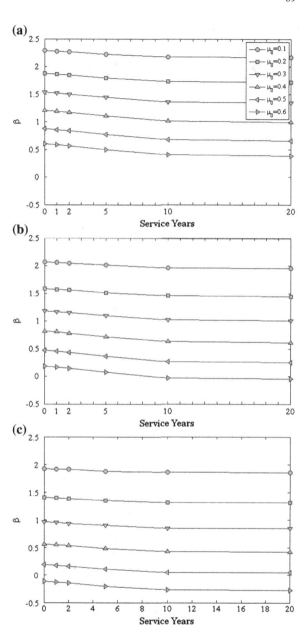

Fig. 6.4 Some results of
reliability index for Case 2.
a $q_f = 200\,\text{MJ/m}^2$,
b $q_f = 300\,\text{MJ/m}^2$,
c $q_f = 400\,\text{MJ/m}^2$

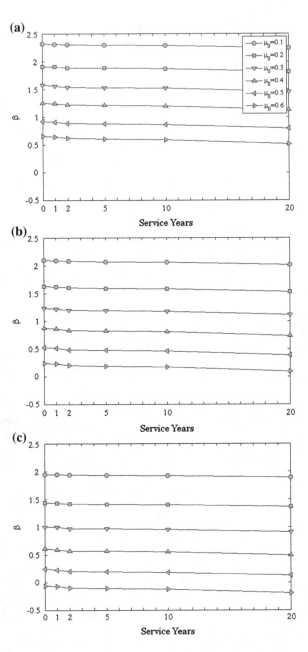

Fig. 6.5 Some results of
reliability index for Case 3.
a $q_f = 200\,\text{MJ/m}^2$,
b $q_f = 300\,\text{MJ/m}^2$,
c $q_f = 500\,\text{MJ/m}^2$

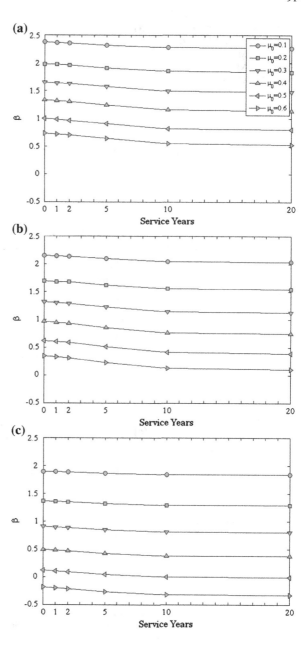

Fig. 6.6 Some results of
reliability index for Case 4.
a $q_f = 200\,\text{MJ/m}^2$,
b $q_f = 300\,\text{MJ/m}^2$,
c $q_f = 500\,\text{MJ/m}^2$

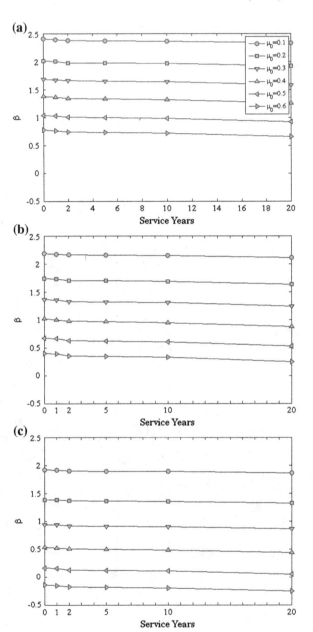

Fig. 6.7 Some results of
reliability index for Case 5.
a $q_f = 300\,\mathrm{MJ/m^2}$,
b $q_f = 400\,\mathrm{MJ/m^2}$,
c $q_f = 600\,\mathrm{MJ/m^2}$

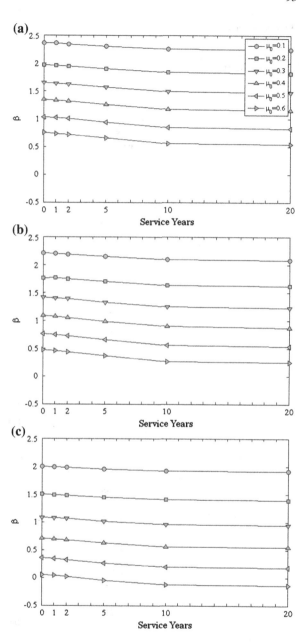

Fig. 6.8 Some results of
reliability index for Case 6.
a $q_f = 300 \, \text{MJ/m}^2$,
b $q_f = 400 \, \text{MJ/m}^2$,
c $q_f = 600 \, \text{MJ/m}^2$

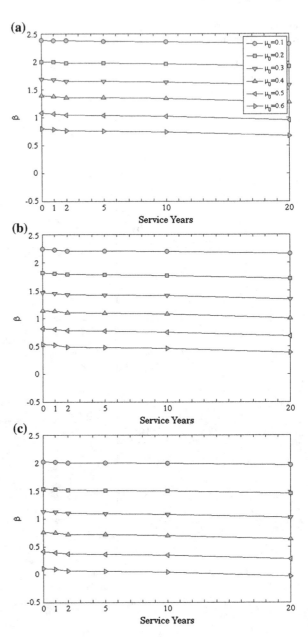

Fig. 6.9 Some results of reliability index for Case 7. **a** $q_f = 200\,\text{MJ/m}^2$, **b** $q_f = 300\,\text{MJ/m}^2$, **c** $q_f = 400\,\text{MJ/m}^2$

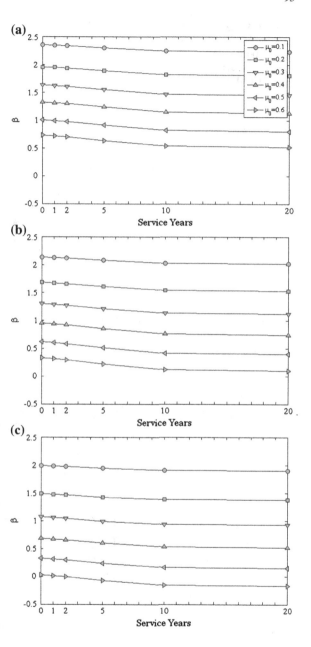

Fig. 6.10 Some results of
reliability index for Case 8.
a $q_f = 200\,\text{MJ/m}^2$,
b $q_f = 300\,\text{MJ/m}^2$,
c $q_f = 400\,\text{MJ/m}^2$

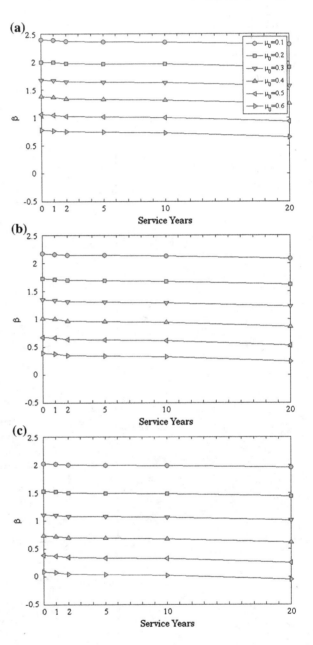

Fig. 6.11 Some results of
reliability index for Case 9.
a $q_f = 200\,\text{MJ/m}^2$,
b $q_f = 300\,\text{MJ/m}^2$,
c $q_f = 500\,\text{MJ/m}^2$

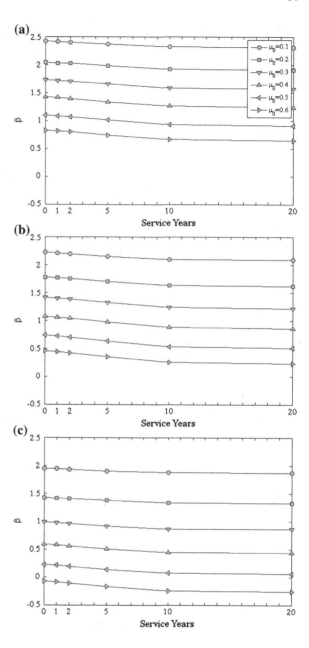

Fig. 6.12 Some results of
reliability index for Case 10.
a $q_f = 200\,\text{MJ/m}^2$,
b $q_f = 300\,\text{MJ/m}^2$,
c $q_f = 500\,\text{MJ/m}^2$

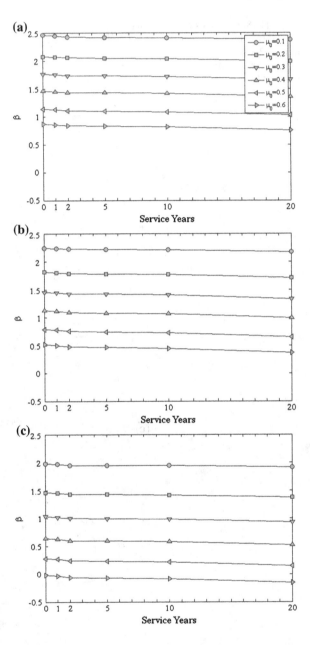

Fig. 6.13 Some results of reliability index for Case 11. **a** $q_f = 300\,\text{MJ/m}^2$, **b** $q_f = 400\,\text{MJ/m}^2$, **c** $q_f = 600\,\text{MJ/m}^2$

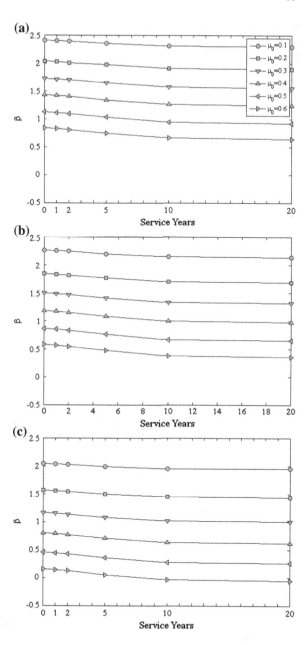

Fig. 6.14 Some results of
reliability index for Case 12.
a $q_f = 300\,\mathrm{MJ/m^2}$,
b $q_f = 400\,\mathrm{MJ/m^2}$,
c $q_f = 600\,\mathrm{MJ/m^2}$

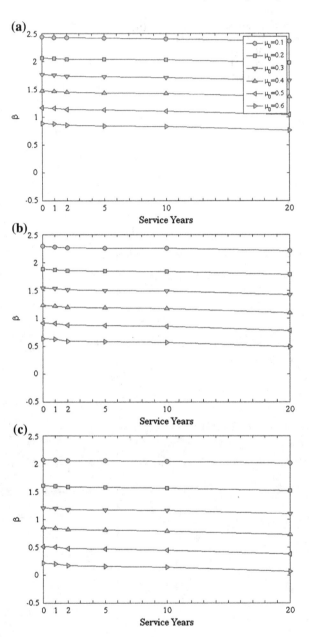

Fig. 6.15 Difference among
reliability index for 0 and 20
service years with respect to
μ_0

Fig. 6.16 Difference among
reliability index for 0 and 20
service years with respect to
β at 0 aging year

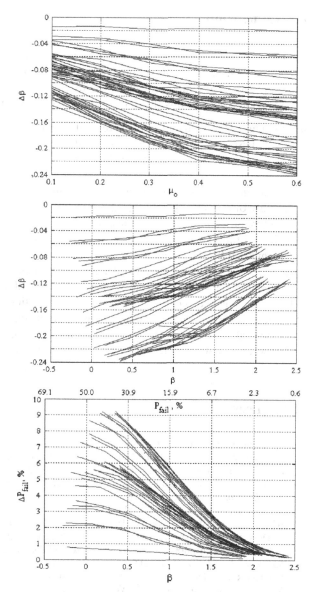

Fig. 6.17 Difference among
failure probability for 0 and
20 service years with respect
to β at 0 aging year

6.6 Conclusion

Intumescent coatings are now the dominant passive fire protection materials used in engineering. Due to its organic components, intumescent coating has aging problems. The reliability of structures protected by intumescent coatings should be investigated. Based on the results of this study, the following conclusions can be drawn:

- Aging has effect of decreasing the reliability index, β, of steel columns protected by intumescent coatings. That decrease effect increases with aging year increases. The decrease of β is more serious for cases with high μ_0 than for cases with low μ_0.
- For the investigated cases with low load ratio ($\mu_0 \leq 0.3$), the amount of the decrease of β due to aging effect, $\Delta\beta$, is less than 0.2 (the corresponding increase in failure probability, ΔP_{fail}, is less than 3 %); and for the investigated cases with high load ratio ($\mu_0 > 0.3$), the maximum $\Delta\beta$ is about 0.24 (the corresponding maximum ΔP_{fail} is about 9 %).
- Limited by test data on thermal insulation properties of intumescent coating with aging effects, the study has only considered steel columns protected by two different thickness of coatings. Also, many assumptions have been made on the coating properties. Further studies using different coatings with various thickness are needed to obtain a general conclusion on the aging effect on reliability of intumescent coating protected steel columns in fire conditions.

References

1. S. Melink, Fire Saf. J. **20**, 83 (1993)
2. C. Thauvoye, B. Zhao, J. Klein, M. Fontana, in *Fire Safety Science—Proceedings of the Ninth International Symposium*, Singapore, 2008, pp. 991–1002
3. E. Coal, S.C. (ECSC), Natural fire safety concept, valorization project. Technical Report (2001)
4. BSI, *Eurocode 1: Actions on Structures—Part 1–2: General Rules—Actions on Structures Exposed to Fire* (British Standard, 2002)
5. JCSS, JCSS probabilistic model code, part II—load models. Technical Report (2001)
6. D. Drysdale, An Introduction to Fire Dynamics, 2nd edn. (Wiley, New York, 1999)
7. S. Iqbal, R. Harichandran, in *Proceedings of the 10th International Conference on Structural Safety and Reliability* (Osaka, 2009)
8. S. Iqbal, R. Harichandran, Fire Saf. J. **46**, 234 (2011)
9. M. Lemaire, Structural reliability (Wiley, New York, 2009)
10. B. Ellingwood, J. Fire Prot. Eng. **14**, 43 (2005)
11. L. Wang, Y. Wang, G. Li, in *Proceedings of the Sixth International Conference on Structures in Fire*, MI, 2010, pp. 735–42
12. G. Li, C. Zhang, G. Lou, Y. Wang, L. Wang, Fire Technol. (2011)

Chapter 7
Service Life of Intumescent Coatings

7.1 Introduction

This chapter presents a probabilistic approach to determine the service life of intumescent coatings for steel columns. A example is provided to show the application of the approach.

7.2 Probability of Fire Ignition

The probability of fire occurrences in a given building or area depends upon the number and type of ignition sources present. This varies during any day and over a period of time and is a strong function of human activity. In stochastic modeling, fire occurrences are always assumed as random point events in time according to the Poisson process [1]. Thus the probability of the occurrence of x fires in time interval t within a particular building is given by

$$P(X = x) = \frac{1}{x!}\lambda t^x e^{-\lambda t} \tag{7.1}$$

where X is the number of fire occurrences in time interval t; λ is the mean fire ignition rate or the average number of fire occurrences per unit time interval.

In general the probability of fire occurrences in buildings that are divided into fire resisting compartments or storeys, increases with the size of the building. For a uniformly compartmented building (each compartment in the building has the same floor area, is equally equipped, and is used in the same way), the rate of fire ignition per unit time interval, λ, can be calculated by

$$\lambda = A_F \lambda_p \tag{7.2}$$

© Springer-Verlag Berlin Heidelberg 2015
C. Zhang, *Reliability of Steel Columns Protected by Intumescent Coatings Subjected to Natural Fires*, Springer Theses, DOI 10.1007/978-3-662-46379-6_7

Table 7.1 Values for λ_p
given in JCSS [2]

Type of building	λ_p
Dwelling/school	$0.5\text{--}4 \times 10^{-6}$
Shop/office	1×10^{-6}
Industrial building	$2\text{--}10 \times 10^{-6}$

where A_F is the floor area of the building; and λ_p is the rate of the fire ignition per unit floor area per unit time interval.

In practice, only a few buildings such as offices, apartments and hotels can be approximated as being uniformly compartmented. Most buildings are multi-purpose buildings which have unequal compartments or various sectors. A multi-purpose building can be subdivided into several different sectors, each of which is composed of a number of equal compartments. Let Y_i be the number of fire occurrences in time interval t within sector i, the number of fire occurrences within the building is the sum of Y_i, $\sum Y_i$. Consider $Y_1, Y_2, \ldots$, are all independent Poisson random variables, the distribution of $\sum Y_i$ is Poisson distribution with parameter $\sum \lambda_i$, where λ_i is the fire ignition rate per unit time interval in sector i. $\sum Y_i$ is the same as X in Eq. 7.1 and $\sum \lambda_i$ is the same as λ in Eq. 7.1.

Table 7.1 gives the values for the rate of the fire ignition per unit floor area per unit time interval λ_p for different type of buildings in JCSS [2].

7.3 Probability of Flashover Occurrence

From the structural point of view, the fully developed or post-flashover fires are likely to cause structures fail. The probability of flashover occurrence, $P\,(flashover)$, is calculated by

$$P\,(flashover) = P\,(flashover|ignition) \times P\,(ignition) \tag{7.3}$$

where $P\,(flashover|ignition)$ is the probability of flashover for given ignition, depending on the type of active protection measures, and $P\,(ignition)$ is the probability of ignition. Table 7.2 gives the values for the probability of flashover for given ignition in JCSS [2].

7.4 Probability of Structural Failure by Fire

The probability of structural failure by fire, $P\,(fail)$, is calculated by

$$P\,(fail) = P\,(fail|flashover) \times P\,(flashover) \tag{7.4}$$

Table 7.2   Values for $P(\textit{flashover}|\textit{ignition})$ given in JCSS [2]

Protection method	$P(\textit{flashover}|\textit{ignition})$
Public fire brigade	10^{-1}
Sprinkler	10^{-2}
High standard fire brigade on site, combined with alarm system (industries only)	10^{-3}–10^{-2}
Both sprinkler and high standard residential fire brigade	10^{-4}

where $P(\textit{fail}|\textit{flashover})$ is the probability of structural failure in post-flashover fire, determined by reliability analysis as given in Sect. 8.3.2.

7.5 Service Life of Intumescent Coating

The probability of structural failure by fire should not exceed a target probability. In Eurocode EN 1990 [3], the target probability for structural fire design is 7.23×10^{-5} (the corresponding reliability index β is 3.8). The service life of intumescent coating can be derived from

$$P(\textit{fail}) = P(\textit{target}) \tag{7.5}$$

where $P(\textit{target})$ is the target probability for structural fire design.

7.6 An Example

Problem   A office building is uniformly compartmented with Comp 1 in Table 6.2. The floor area of the building is $A_f = 24,000\,\text{m}^2$. The design fire load density is $q_f = 300\,\text{MJ/m}^2$, and the load ratio is $\mu_0 = 0.1$. The active fire protection measure is by sprinkler. The design service life for the building is 50 years. The steel column and intumescent coating for Case 1 in Table 6.2 are used. Assess the service life of the intumescent coating.

Solution:

1. Get λ_p from Table 7.1: $\lambda_p = 1 \times 10^{-6}\,\text{m}^{-2}\,\text{year}^{-1}$.
2. Calculate λ from Eq. 7.2: $\lambda = A_F \lambda_p = 24,000 \times 1 \times 10^{-6} = 2.4 \times 10^{-2}$/year.
3. From Table 7.2, for sprinkler protected building: $P(\textit{flashover}|\textit{ignition}) = 10^{-2}$;
4. By Eq. 7.3, the probability of the occurrence of one post-flashover fire is:

- for time interval of 1 year: $P(1, 1) = \frac{2.4 \times 10^{-2} \times 1^1}{1!} e^{-2.4 \times 10^{-2} \times 1} \times 10^{-2} = 2.34309 \times 10^{-4}$;

- for time interval of 10 years: $P(1, 10) = \frac{2.4 \times 10^{-2} \times 10^{1}}{1!} e^{-2.4 \times 10^{-2} \times 10} \times 10^{-2} =$ 1.887907×10^{-3};
- for time interval of 20 years: $P(1, 20) = \frac{2.4 \times 10^{-2} \times 20^{1}}{1!} e^{-2.4 \times 10^{-2} \times 20} \times 10^{-2} =$ 2.97016×10^{-3};
- for time interval of 50 years: $P(1, 50) = \frac{2.4 \times 10^{-2} \times 50^{1}}{1!} e^{-2.4 \times 10^{-2} \times 50} \times 10^{-2} =$ 3.614331×10^{-3}.

5. From Chap. 6, the failure probability of steel column protected by intumescent coating is

   - for 0 service year: $P(fail|flashover, 0) = 0.0191$;
   - for 1 service year: $P(fail|flashover, 1) = 0.0198$;
   - for 10 service year: $P(fail|flashover, 10) = 0.0249$;
   - for 20 service year: $P(fail|flashover, 20) = 0.0253$.

6. If not considering aging effect of intumescent coating, the failure probability of protected steel column by fire in the service life of 50 years is: $P(fail, 50) = 0.0191 \times 3.614331 \times 10^{-3} = 6.90337 \times 10^{-5} < P(target) = 7.23 \times 10^{-5}$, (Safe).

7. If considering aging effect of intumescent coating, the failure probabilities of protected steel column by fire for different service years are:

   - for 1 service year: $P(fail, 1) = 0.0198 \times 2.34309 \times 10^{-4} = 4.63931 \times 10^{-6} < 7.23 \times 10^{-5}$, (Safe);
   - for 10 service years: $P(fail, 10) = 0.0249 \times 1.887907 \times 10^{-3} = 4.70089 \times 10^{-5} < 7.23 \times 10^{-5}$, (Safe);
   - for 20 service years: $P(fail, 20) = 0.0253 \times 2.97016 \times 10^{-3} = 7.51451 \times 10^{-5} > 7.23 \times 10^{-5}$, (Unsafe).

Answer If not considering aging effect of intumescent coating, in the service life of the building (50 years), the failure probability of the protected column by fire is less than the target probability; however, in practice, due to aging effect of intumescent coating, after 20 service years, the failure probability of protected column will be greater than the target probability. Therefore, the service life for the intumescent coating for this example is about 20 years.

References

1. T. Lie, *Technical Paper* (Division of Building Research, National Research Council Canada, 1974)
2. JCSS, Jcss probabilistic model code, part ii–load models. Technical report (2001)
3. EN 1990, in *Eurocode—Basis of structural design* (British Standards Institution (BSI), 2002)

Chapter 8
Conclusions and Further Work

8.1 Introduction

This research is diverse and wide ranging. It has covered in some detail all areas necessary to determine the reliability of steel columns protected by intumescent coatings subjected to natural fires. One zone model for predicting the gas temperature of post-flashover fires and its sub-models have been reviewed in detail. The concept of equivalent fire exposure and different approaches to calculate the equivalent fire exposure time have been reviewed. One dimensional condensed heat transfer model for predicting steel temperature of insulated steel members and different formulae derived from this model have been reviewed. Shortcomings of current formulae for predicting steel temperature of insulated steel members has been investigated and a simple approach has been proposed to calculate the maximum steel temperature of insulated steel members subjected to natural fires. The behavior of intumescent coatings under heating has been studied and mathematical heat transfer model for intumescent coatings has been reviewed. Current procedures for assessing the thermal insulation property of intumescent coatings has been reviewed and their shortcomings are presented. A simple procedure has been proposed to assess fire resistance of intumescent coatings by using the concept of equivalent constant thermal resistance. Current study on aging effect on thermal insulation property of intumescent coatings has been reviewed and the values of equivalent constant thermal resistance of intumescent coatings with different aging years have been calculated. The fundamental behavior of steel columns under fire conditions and experimental studies on steel columns in fire have been reviewed. The accuracy and limitation of calculation approaches to predict the buckling and limiting temperature of steel columns have been investigated by comparing with test data reported in literature. The model errors or professional factors for calculation approaches have been characterized. The statistics of parameters in reliability analysis of steel columns protected by intumescent coatings to natural fires have been determined from literature. Basic of reliability theory has been presented and Monte Carlo simulations have been conducted to obtain the failure probability and reliability index. A probabilistic approach to determine

C. Zhang, *Reliability of Steel Columns Protected by Intumescent Coatings Subjected to Natural Fires*, Springer Theses, DOI 10.1007/978-3-662-46379-6_8

the service life of intumescent coatings for steel columns has been provided. The
purpose of this chapter is to summarize and appraise the conclusions of this research.
Suggestions for possible further work are also made.

8.2 Summary and Conclusions

Some of the key summary points and conclusions raised by this work are listed
below:

- A modified one zone model has been proposed to predict the mean temperature
  of steel members subjected to post-flashover fires. In the model, a quantity which
  considers the heat sink effect of steel members in the fire compartment is added
  to the heat balance equation for the traditional one zone compartment fire model.
  Based on the modified one zone model, numerical or analytical approaches can
  be developed to give more rational prediction of steel and gas temperatures in
  post-flashover fire conditions.
- Current formulae provided by fire codes in different countries for calculating the
  steel temperature of insulated steel members in fire conditions are based on the standard fire model, which might give unacceptable results for calculation in natural
  fires. Besides, when using those formulae, iterative computations should be always
  processed, which is not convenient for daily design works and is not efficient for
  reliability analysis which usually includes hundreds of thousands of simulation
  loops for a single case.
- A simple approach has been developed for calculating maximum steel temperature
  of insulated steel members in natural fires. The approach adopts time equivalent to
  relate natural fires with the standard fire, and use a simple quadratic equation for
  calculating the maximum steel temperatures. By comparing with numerical results
  and test data, the proposed approach can give satisfactory prediction of maximum
  steel temperatures in the range from 300 to 600 °C. The professor factor of the
  approach has been characterized by test data, which has a mean of 0.955 and a
  COV of 0.014, and can be best described by lognormal distribution. The approach
  only needs hand calculations which is easy and convenient for practical usage, and
  the approach is given in a closed form which is efficient for reliability analysis.
- Intumescent coatings will react at high temperatures and the thermal properties
  of intumescent coatings can not be measured directly by the current standard
  test methods which are originally designed for the traditional inert fireproofing
  materials. A simple procedure has been proposed to assess the fire resistance of
  intumescent coatings by using the concept of equivalent constant thermal resistance. By using the equivalent constant thermal resistance derived at a critical steel
  temperature of 550 °C, the calculated steel temperatures agree well with the test
  data in the range of the concerned limiting temperatures from 400 to 600 °C.
- In practice, when specifying coating fire protection for steel structure, it assumes
  that the coating is correctly applied and its performance meets the fire protection

needs without degradation over time. However, since the organic components of intumescent coating, it should be expected that the fire protection function of intumescent coating over time would not be as reliable as when freshly applied. The values of constant thermal resistance of two thickness of intumescent coating exposed to aging condition reported in literature are derived. It shows that aging has effect of reducing the constant thermal resistance of intumescent coatings.

- In current Eurocode, the buckling temperature of steel columns can be calculated by either using an analytical approach or using a simple closed-form equation. The accuracy and limitations of those two calculation approaches have been investigated by comparing with test data on steel columns at elevated temperatures reported in literature. The two approaches are found to give acceptable prediction for tests with moderate utilization factor, give unacceptable prediction for tests with either high utilization factor or low utilization factor. The professional factors for the two approaches have also been characterized. The professional factor for the simple equation has a mean of 0.949 and a COV of 0.016, and can be best described by a extreme value distribution; and the professional factor for the analytical approach has a mean of 1.018 and a COV of 0.013, and can be well described by either a normal, gamma or lognormal distribution.

- The accuracy and limitations of simple calculation approaches for predicting the buckling or critical temperature of axially restrained steel columns have also been investigated by comparing with test data on axially restrained steel columns at elevated temperatures reported in literature. The simple approach based on the Eurocode 3 gives unacceptable prediction of buckling temperatures for axially restrained steel columns. The approach proposed by Neves et al. gives acceptable prediction of limiting temperature of axially restrained steel components, and the professional factor of the approach has a mean of 0.949 and a COV of 0.006, and can be best described by normal distribution.

- Reliability of steel columns protected by intumescent coating subjected to natural fires has been investigated. Particularly, Monte Carlo simulations are conducted to investigate the aging effect on failure probability and reliability index of the column. Aging has effect of decreasing the reliability index, β, of steel columns protected by intumescent coatings. That decrease effect increases with aging year increases. The decrease of β is more serious for cases with high load ratio μ_0 than for cases with low μ_0. For the investigated cases with low load ratio ($\mu_0 \leq 0.3$) the amount of the increase of β due to aging effect, $\Delta\beta$, is less than 0.2 (the corresponding increase in failure probability, ΔP_{fail}, is less than 3 %); and for the investigated cases with high load ratio ($\mu_0 > 0.3$), the maximum $\Delta\beta$ is about 0.24 (the corresponding maximum ΔP_{fail} is about 9 %).

- A probabilistic approach has been provided to determine the service life of intumescent coatings for steel columns. The approach compares the failure probability of the protected steel columns with the target probability of the structural fire design. The probability of fire occurrence and the probability of flashover are considered and determined from codes. An example is given to determine the service life of ICs for protecting steel columns in an office building. The service life for the intumescent coatings in the example is 20 years. The approach is also applicable for designing the traditional inert fire proofing materials.

8.3 Further Work

The last 20 years has been the development of fire modeling, structural behavior in fire, and performance of intumescent coating under heating increase at a rapid rate with their use today commonplace in daily fire safety design. Although the research described in this thesis has helped to develop the understanding regarding reliability of steel structures to fire, there are inevitable many issues that remain inadequately or as yet, totally unresolved.

8.3.1 Testings

- Due to its complexity, the performance of intumescent coatings under heating has not been studied well. More testing on performance of intumescent coatings in both standard fire and natural fires would help to develop the understanding of the performance and lead to a greater source of data which validation of theoretical models can be carried out.
- The thermal insulation property of different type of intumescent coatings might be different significantly. As a result, although a number of studies on intumescent coatings have been reported in literature, available data to characterized the thermal insulation property of intumescent coatings is limited. Increased testing of the constant thermal resistance of intumescent coatings with different thickness would help to obtain the statistics of the constant thermal resistance of intumescent coatings.
- Only recently, the aging problem of intumescent coatings has been concerned in structural fire research. Few work on this issue has been done. Increased testing of thermal insulation property of intumescent coatings with aging effects are required to obtain a general understanding on aging of intumescent coatings.
- Test data on temperature of insulated steel members in natural fires are also limited. Increased testing of insulated steel members with different section factors and various fire protections in different natural fires would lead to a greater source of data which more validation exercise can be carried out.
- More test data on limiting temperature of restrained steel columns in fire are required to further validate the current simple approach or to develop new approaches. Then, robust behavior of steel columns coupled in global structures can be considered in reliability analysis.

8.3.2 Reliability Analysis

Limited by test data on thermal insulation properties of intumescent coating with aging effects, the current study has only considered steel columns protected by two different thickness of coatings. Also, many assumptions have been made on the coating properties. Further studies using different coatings with various thickness are needed to obtain a general conclusion on the aging effect on reliability of intumescent coating protected steel columns in fire conditions.

Appendix A
Fundamentals of Heat Transfer

A.1 Heat Conduction

When a temperature gradient exists in a body, energy will transfer from the high-temperature region to the low-temperature region by conduction. The general equations of heat conduction is given by [1]

$$\nabla \cdot (k\nabla T) + \dot{q}_{in} = \rho c \frac{\partial T}{\partial t} \tag{A.1}$$

where T is temperature; $\dot{q}_{in}$ is energy generated per unit volume; k, ρ, and c are thermal conductivity, density and specific heat, respectively.

In fire engineering, one-dimensional (1D) heat transfer is usually considered. The equation of 1D conduction is given by (ignore energy generation, thus $\dot{q}_{in} = 0$)

$$\frac{\partial^2 T}{\partial x^2} = \frac{1}{\alpha} \frac{\partial T}{\partial t} \tag{A.2}$$

where $\alpha = k / \rho c$ is thermal diffusivity.

Fourier's law states that the quantity of heat transferred per unit time per unit area is proportional to the temperature gradient, thus

$$\dot{q} = -k \frac{\partial T}{\partial x} \tag{A.3}$$

To solve Eq. A.2, the following boundary conditions are usually used [2, 3]

1. Initial condition,

$$T(x, 0) = T_\infty \tag{A.4}$$

2. Dirichlet condition,

$$T(0, t) = T_g(t) \tag{A.5}$$

© Springer-Verlag Berlin Heidelberg 2015
C. Zhang, *Reliability of Steel Columns Protected by Intumescent Coatings Subjected to Natural Fires*, Springer Theses, DOI 10.1007/978-3-662-46379-6

3. Neumann condition,

$$-k\frac{\partial T}{\partial x}(0, t) = \dot{q}_0 = h[T_g(t) - T(0, t)] \qquad (A.6)$$

where, x denotes opposite normal direction to the surface; T_∞, T_g are environment temperature and fire temperature, respectively; $\dot{q}_0$ is heat flux per unit time transferred from fire environment to the surface; and h is coefficient for convection or radiation.

In unsteady or transient conduction, Boit number is used to determine whether a body can be treated as thermally thin or thermally thick, which is defined by [1]

$$Bi = \frac{\delta/k}{1/h} = \frac{h\delta}{k} \qquad (A.7)$$

where, $\delta = V/A$ is the characteristic length of the body, in which V and A are the body's solid volume and its surface area.

In engineering calculation, if $Bi < 0.1$ the body is treated as thermally thin and if $Bi > 0.1$ the body is treated as thermally thick. For thermally thin body, the temperature gradient within the body can be ignored and lump-capacity method can be applied to temperature calculation; and for thermally thick body, the temperature gradient throughout the body can not be ignored, as demonstrated in Fig. A.1. In the figure, a plane wall initially at a uniform temperature T_i experiences heating when

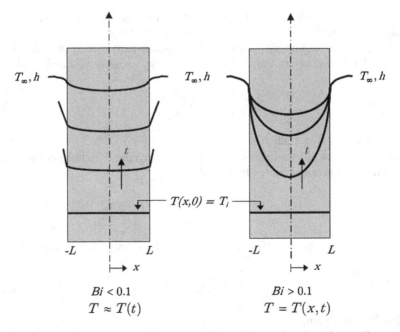

Fig. A.1 Transient temperature distributions for different Biot numbers in a plane wall symmetrically cooled by convection

immersed in a fluid of $T_\infty < T_i$. The problem can be treated as one dimensional in x and the temperature variation with position and time, $T(x, t)$ is shown. The temperature variation throughout the section is seen to a strong function of the Biot number. When the Biot number is small, temperature gradients in the solid are small and the main temperature difference is between the solid and the fluid, and the solid temperature remains nearly uniform as it increases to T_∞. For large values of the Biot number, temperature gradients within the solid can be significant and the temperature difference across the solid is much larger than that between the surface and the fluid.

The temperature of a thermally thin body ($Bi < 0.1$) under a initial boundary given by Eq. A.4 and a Neumann boundary given by Eq. A.6 (A example of this case is a bare steel member fully engulfed by fire), calculated by using lumped-capacity method, is given by [1]

$$\frac{T - T_\infty}{T_g - T_\infty} = exp\left(-\frac{hA}{\rho cV}t\right) = e^{-Bi \cdot Fo} \tag{A.8}$$

where $Fo = \alpha t / \delta^2$ is Fourier number. Take $Fo = 1$, we can get the thermal penetration time of the thermal thin body, given by [4]

$$t_{pthin} = \frac{\delta^2}{\alpha}. \tag{A.9}$$

The thermal penetration time is the time required for a thermal pulse to reach the back face of the body [5].

For thermally thick bodies ($Bi > 0.1$) like concrete walls and concrete slabs, in fire engineering calculations the bodies are usually treated as semi-infinite bodies. The temperature of a semi-infinity solid under a initial boundary given by Eq. A.4 and a Neumann boundary given by Eq. A.6 is give by [2, 3]

$$\frac{T - T_\infty}{T_g - T_\infty} = erfc\left(\frac{x}{2\sqrt{\alpha t}}\right) - exp\left(\frac{xh}{k} + \frac{\alpha t}{(k/h)^2}\right) \cdot erfc\left(\frac{x}{2\sqrt{\alpha t}} + \frac{\sqrt{\alpha t}}{k/h}\right) \tag{A.10}$$

where, $erfc(\xi) = 1 - erf(\xi)$. $erf(\xi)$ is the error function, given by

$$erf(\xi) = \frac{2}{\pi}\int_0^\xi e^{-\eta^2}d\eta \tag{A.11}$$

Take $\alpha t / (k/h)^2 = 1$ or $\sqrt{\alpha t}/(k/h) = 1$, we get the thermal penetration time of the thermally thick body, given by [4]:

$$t_{pthick} = \frac{(k/h)^2}{\alpha} = \frac{k\rho c}{h^2} \tag{A.12}$$

The temperature of the same semi-infinity body under a initial boundary given by Eq. A.4 and a Dirichlet boundary given by Eq. A.5 is given by

$$\frac{T - T_\infty}{T_g - T_\infty} = 1 - erf\left(\frac{x}{2\sqrt{\alpha t}}\right) \qquad (A.13)$$

Take Eq. A.13 = 0.005, we get

$$x = L \approx 4\sqrt{\alpha t} \qquad (A.14)$$

where L is the distance to the surface at which the temperature decreases to 0.5 % of the surface temperature. Equation A.14 indicated that a wall or slab, can be treated as a semi-infinite solid with little error, provided its thickness is greater than $4\sqrt{\alpha t}$. In fire engineering, if the thickness of a wall or slab is greater than $2\sqrt{\alpha t}$, the semi-infinite solid assumption is used and the corresponding thermal penetration time is given by [2, 5]

$$t_p = \frac{1}{\alpha}\left(\frac{\delta_w}{2}\right)^2 \qquad (A.15)$$

where δ_w is the thickness of the wall or slab.

A.2 Heat Convection

Convection describes the energy transfer between a surface and a fluid moving over that surface as a result of an imposed temperature difference. Strictly, convection is not a basic model of heat transfer, rather it can be considered as a combined effect of conduction and the motion of some transmitting medium (The only two basic models of heat transfer are conduction and radiation). In general, however, convection is treated as a separate mode of heat transfer involving complex relationships between velocity, temperature and concentration distributions.

Newton's law of cooling states that the heat flux transferred by convection, $\dot{q}_c$, is proportional to the difference in temperature between the surface and the fluid, that

$$\dot{q}_c = h_c(T_\infty - T_s) \qquad (A.16)$$

where the proportionality constant h_c is the convection heat transfer coefficient or film coefficient; T_s is surface temperature; and T_∞ is the fluid temperature far away from the surface.

By the boundary layer concept, the heat flux transferred from fluid to surface can also be calculated by Fourier's law in conduction, that [6]

$$\dot{q} = -k_f \left.\frac{\partial T}{\partial y}\right|_{y=0} \approx \frac{k_f}{\delta_\theta}(T_\infty - T_s). \qquad (A.17)$$

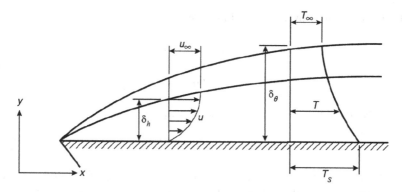

Fig. A.2 Velocity and thermal boundary layers on a flat plate [94]

From Eq. A.16 equal to Eq. A.17, we get

$$h_c = \frac{-k_f \cdot \partial T / \partial y \mid_{y=0}}{T_\infty - T_s} \approx \frac{k_f}{\delta_\theta} \qquad (A.18)$$

where k_f is the conductivity of the fluid; and δ_θ is the thickness of thermal boundary layer, as illustrated in Fig. A.2. Figure A.2 considers a free stream fluid with velocity u_∞ flow over a flat plate. The particles in contact with the plate surface at zero velocity retard the motion of the particles in the adjacent layer and so on, until at some distance $y = \delta_h$ from the surface the effect becomes negligible. The quantity δ_h represents the velocity boundary layer thickness and is defined as the value of y for which $u = 0.99u_\infty$. Just as a velocity boundary layer develops because of viscous effects near the surface, a thermal boundary layer develops due to heat transfer between the free stream and the surface if their temperatures are different. The thermal boundary thickness, δ_θ, is defined as the value of y for which the ratio $[(T - T_s)/(T_\infty - T_s)] = 0.99$ [3, 4].

Equation A.18 can be alternatively expressed in the following form [6]

$$\frac{\partial \left(\frac{T-T_s}{T_\infty - T_s} \right)}{\partial (y/L)} \Bigg|_{y/L=0} = \frac{h_c L}{k_f} = Nu \qquad (A.19)$$

where L is the characteristic dimension of the surface, e.g. the length of the plate in Fig. A.2; Nu is Nusselt number. For flow over a horizontal flat plate [2],

$$Nu = 0.332 Re^{1/2} Pr^{1/3} \qquad (A.20)$$

provided the flow is laminar and for turbulent flow

$$Nu = 0.037 Re^{4/5} Pr^{1/3}; \qquad (A.21)$$

and for buoyancy flow over a vertical flat plate

$$Nu = 0.59(G_r \cdot P_r)^{1/4} \tag{A.22}$$

$$Nu = 0.59(G_r \cdot P_r)^{1/4} \tag{A.23}$$

provided that the flow is laminar; and for turbulent flow

$$Nu = 0.13(G_r \cdot P_r)^{1/3} \tag{A.24}$$

where $Re = \rho u_\infty L/\mu$, $Pr = \nu/\alpha = \mu/\rho\alpha$ and $G_r = g\beta(T_\infty - T_s)L^3/\nu^2$ are Reynold number, Prandtl number and Grashof number, respectively. ρ, u_∞ and μ are density, velocity and dynamic viscosity of the fluid, respectively and β is the reciprocal of 273 K.

In natural or free convection, the flow is created by buoyancy induced by the temperature difference between the boundary layer and the ambient fluid (In forced convection, the fluid is flowing as a continuous stream past the surface), and for a vertical flat plate,

$$Nu = 0.59(G_r \cdot P_r)^{1/4} \tag{A.25}$$

provided that the flow is laminar; and for turbulent flow

$$Nu = 0.13(G_r \cdot P_r)^{1/3} \tag{A.26}$$

where $Re = \rho u_\infty L/\mu$, $Pr = \nu/\alpha = \mu/\rho\alpha$ and $G_r = g\beta(T_\infty - T_s)L^3/\nu^2$ are Reynold number, Prandtl number and Grashof number, respectively. ρ, u_∞ and μ are density, velocity and dynamic viscosity of the fluid, respectivelyand β is the reciprocal of 273 K. For flow over a flat plate, when the Relynold number reaches the critical value of about 5×10^5, the flow changes from laminar to turbulent [4], as shown in Fig. A.3; and for flow over a vertical flat plate, when the Rayleigh number $Ra = G_r \cdot P_r$ reaches the critical value of about 10^9 the flow changes from laminar to turbulent [2].

Typically, h_c takes values in the range 5–50 W/m^{2}K and 25–250 W/m^{2}K for natural convection and forced convection in air, respectively [2]. In natural or free convection, fluid motion generated by buoyancy induced by temperature gradients in the fluid, whilst in forced convection, fluid motion is generated mechanically through the use of a fan, blower, nozzle, jet, etc..

A.3 Heat Radiation

Because of their temperature, all bodies constantly emit energy by a process of electromagnetic radiation; we refer to this as thermal radiation. The wavelengths for thermal radiation are in the range 10^{-1}–$10^3 \mu$m. By Planck's distribution law, the

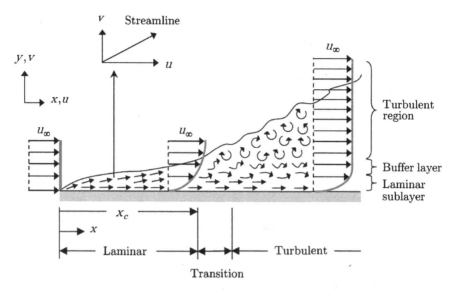

Fig. A.3 Velocity boundary layer development on a flat plate showing laminar and turbulent regions, $Re_{x_c} \approx 5 \times 10^5$ [95]

spectral (or monochromatic) intensity of blackbody radiation is given by

$$E_{b,\lambda} = \frac{2\pi c^2 h \lambda^{-5}}{exp(ch/kT) - 1} \tag{A.27}$$

where c is the velocity of light; h is Planck's constant; k is Boltzmann's constant; and T is the absolute temperature. The total emissive power of a black body is

$$E_b = \int_0^\infty E_{b,\lambda} d\lambda = \sigma T^4 \tag{A.28}$$

where $\sigma = 5.67 \times 10^{-8} \, \mathrm{W/m^2 K^4}$ is Stefan-Boltzmann constant.

A blackbody is defined as an ideal body that allows all the incident radiation to pass into it (no reflected energy) and internally absorbs all the incident radiation (no transmitted energy) [7]. Other types of surfaces do not radiate as much energy as the blackbody. The total energy emitted by a real surface is

$$E = \varepsilon(T_s) E_b(T_s) = \varepsilon(T_s) \sigma T_s^4 \tag{A.29}$$

where $\varepsilon(T_s)$ is emissivity of the surface, which is dependent on temperature. In practice, constant values of emissivity for building materials are used. e.g. the emissivity of concrete and structural steel are taken as 0.7 in EC4 [8].

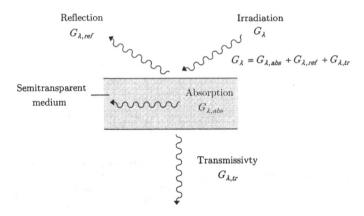

Fig. A.4 Absorption, reflection and transmission processes associated with a semitransparent medium [95]

As shown in Fig. A.4, when a spectral component of radiation strikes a medium surface, portions of this radiation may be reflected, absorbed and transmitted. By energy balance,

$$\rho_\lambda + \alpha_\lambda + \tau_\lambda = 1 \qquad (A.30)$$

where ρ_λ, α_λ, and τ_λ denotes the fraction of energy absorbed by, reflected at, and transmitted through the surface, or are reflectivity, absorptivity and transmissivity, respectively.

Kirchoff's law states that in order to maintain equilibrium, absorptivity and emissivity must be related by

$$\alpha_\lambda = \varepsilon_\lambda \qquad (A.31)$$

A surface whose emittance is the same for all directions is called a diffuse emitter [9]. If the spectral emittance is the same for all wavelength, thus $\varepsilon = \varepsilon_\lambda$, the surface is gray [9].

Consider a opaque gray surface and take E_{ir} as the incident energy by radiation, or irradiation, then the energy leaving the surface, or radiosity, is the combination of surface emission and reflection of irradiation, as shown in Fig. A.5. Thus

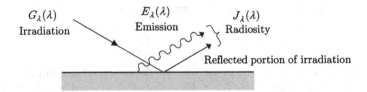

Fig. A.5 Surface radiosity

$$E_{tot} = \rho E_{ir} + \varepsilon E_b = (1 - \alpha)E_{ir} + \varepsilon E_b = (1 - \varepsilon)E_{ir} + \varepsilon E_b \qquad (A.32)$$

Most solids and fluids are opaque bodies, the surface layers of which can absorb all penetrating radiation. For participating medium like gases, intensity of the incoming radiation will reduce with penetration distance by either absorbing or scattering effect of the medium. Consider a monochromatic beam of radiation with intensity $I_\lambda(0)$ passes through a participating medium of thickness L, as shown in Fig. A.6, by Beer's law, the intensity of the radiation beam at point x is given by [6]

$$I_{\lambda 0}(x) = I_\lambda(0)e^{-\rho \kappa_\lambda x} \qquad (A.33)$$

where κ_λ is called the mononchromatic extinction coefficient, which is generally the sum of the absorption coefficient and the scattering coefficient. Correspondingly, for the participating medium of thickness L, the monochromatic absorptance, α_λ, is

$$\alpha_\lambda = \frac{I_\lambda(0) - I_{\lambda 0}(L)}{I_\lambda(0)} = 1 - e^{-\rho \kappa_\lambda L}. \qquad (A.34)$$

Consider Eq. A.31, we get the spectral emissivity for the participating medium of thickness L, that

$$\varepsilon_\lambda = \alpha_\lambda = 1 - e^{-\rho \kappa_\lambda L} \qquad (A.35)$$

where $\rho \kappa_\lambda L$ is called the optical path length or opacity. If $\rho \kappa_\lambda L << 1$, then the medium is optically thin and the medium is transparent to the wavelength λ. However if $\rho \kappa_\lambda L >> 1$ then the medium is optically thick, which implies that the mean penetration distance is much less than the characteristic length L. When this is the case, local radiation results only from local emission.

The outgoing radiation at L in Fig. A.6 is the sum of the reduced penetrating radiation and the emitted radiation by the participating medium, that [4]

$$I_\lambda(L) = I_\lambda(0)e^{-\rho \kappa_\lambda L} + I_{\lambda,b}(1 - e^{-\rho \kappa_\lambda L}) \qquad (A.36)$$

where $I_{\lambda,b}$ is the intensity of blackbody radiation.

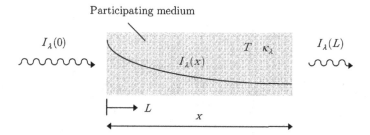

Fig. A.6 Reduction of radiative intensity due to a participating medium

Monatomic and symmetric diatonic molecules such as N_2 and O_2 and H_2 are completely transparent to thermal radiation, while asymmetrical molecules like CO_2, H_2O, CO, and SO_2 absorb (and emit) thermal radiation of certain wavelengths [6]. As a result, the absorptivity (or emissivity) of a air volume is mainly determined by the volume fraction of H_2O and CO_2.

Combustion products include gases and soot. Soot particles are produced as a result of incomplete combustion and are usually observed to be in the form of spheres, agglometrated chunks and long chains. The total emissivity of gas-soot mixture can be calculated approximately by [4]

$$\varepsilon = (1 - e^{-\kappa_m L}) + \varepsilon_g e^{-\kappa_m L} \tag{A.37}$$

where κ_m is the mean absorption coefficient of the mixture, related to parameters like soot volume fraction and temperature [9]

In calculating radiative energy exchange between surfaces, view factor is used to determine how much of the energy that leaves a surface travels toward the surface under consideration. The geometric relations between surfaces are determined by view factor or configuration factor. As shown in A.7, the view factor between two diffuse surfaces A_i and A_j, F_{i-j}, is defined as the portion of all that radiation which leaves surface A_i and strikes surface A_j. Mathematically [9]

$$F_{i-j} = \frac{1}{A_i} \int_{A_i} \int_{A_j} \frac{\cos \theta_i \cos \theta_j}{\pi S_{ij}^2} dA_j dA_i. \tag{A.38}$$

Fig. A.7 View factor

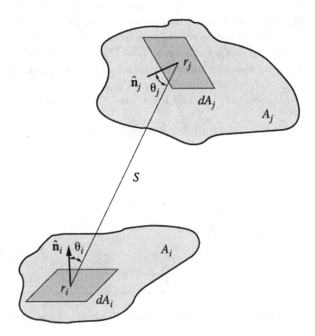

Obviously,

$$A_i F_{i-j} = A_j F_{j-i} \tag{A.39}$$

$$\sum_{j=1}^{N} F_{i-j} = 1. \tag{A.40}$$

Catalog of common view factors can be found in heat transfer handbooks such as [6, 9].

Appendix B
High Temperature Material Properties of Structural Steel

B.1 Thermal Properties

B.1.1 Coefficient of Thermal Elongation

In EC3 [10], the coefficient of elongation of structural steel is given by

$$\alpha_s = 1.2 \times 10^{-5} + 0.8 \times 10^{-8} T_s \quad (20\,^\circ\text{C} < T_s \leq 750\,^\circ\text{C}) \tag{B.1a}$$

$$\alpha_s = 0 \quad (750\,^\circ\text{C} < T_s \leq 860\,^\circ\text{C}) \tag{B.1b}$$

$$\alpha_s = 2 \times 10^{-5} \quad (860\,^\circ\text{C} < T_s \leq 1200\,^\circ\text{C}) \tag{B.1c}$$

In ASCE [11], the coefficient of elongation of structural steel is given by

$$\alpha_s = (0.004 T_s + 12) \times 10^{-6} \quad (20\,^\circ\text{C} < T_s \leq 1000\,^\circ\text{C}) \tag{B.2a}$$

$$\alpha_s = 1.6 \times 10^{-5} \quad (T_s \geq 1000\,^\circ\text{C}) \tag{B.2b}$$

In CECS200 [12], the coefficient of elongation of structural steel is given by

$$\alpha_s = 1.4 \times 10^{-5} \tag{B.3}$$

B.1.2 Specific Heat

In EC3 [10], the specific heat of structural steel is given by

$$c_s = 425 + 0.773 T_s - 0.00169 T_s^2 + 2.22 \times 10^{-6} T_s^3 \quad (20\,^\circ\text{C} \leq T_s < 600\,^\circ\text{C}) \tag{B.4a}$$

© Springer-Verlag Berlin Heidelberg 2015
C. Zhang, *Reliability of Steel Columns Protected by Intumescent Coatings Subjected to Natural Fires*, Springer Theses, DOI 10.1007/978-3-662-46379-6

$$c_s = 666 + \frac{13002}{738 - T_s} \quad (600\,°C \le T_s < 735\,°C) \tag{B.4b}$$

$$c_s = 545 + \frac{17820}{T_s - 731} \quad (735\,°C \le T_s < 900\,°C) \tag{B.4c}$$

$$c_s = 650 \quad (900\,°C \le T_s \le 1200\,°C) \tag{B.4d}$$

In ASCE [11], the specific heat of structural steel is given by

$$c_s = \frac{(0.004T_s + 3.3) \times 10^6}{\rho_s} \quad (20\,°C < T_s \le 650\,°C) \tag{B.5a}$$

$$c_s = \frac{(0.068T_s + 38.3) \times 10^6}{\rho_s} \quad (650\,°C < T_s \le 725\,°C) \tag{B.5b}$$

$$c_s = \frac{(-0.086T_s + 73.35) \times 10^6}{\rho_s} \quad (725\,°C < T_s \le 800\,°C) \tag{B.5c}$$

$$c_s = \frac{4.55 \times 10^6}{\rho_s} \quad (T_s > 800\,°C) \tag{B.5d}$$

where, $\rho_s = 7,850$ kg/m^3 is steel density.

In CECS200 [12], the specific heat of structural steel is given by

$$c_s = 600 \tag{B.6}$$

B.1.3 Thermal Conductivity

In EC3 [10], the thermal conductivity of structural steel is given by

$$k_s = 54 - 0.0333T_s \quad (20\,°C \le T_s < 800\,°C) \tag{B.7a}$$
$$k_s = 27.3 \quad (800\,°C \le T_s \le 1200\,°C) \tag{B.7b}$$

In ASCE [11], the thermal conductivity of structural steel is given by

$$k_s = (-0.022T_s + 48) \quad (20\,°C \le T_s \le 900\,°C) \tag{B.8a}$$
$$k_s = 28.2 \quad (T_s > 900\,°C) \tag{B.8b}$$

In CECS200, the thermal conductivity of structural steel is given by

$$k_s = 45 \tag{B.9}$$

B.2 Structural Properties

B.2.1 Elastic Modulus and Yield Strength

In EC3, the elastic modulus and yield strength of structural steel at elevated temperature are given by

$$E_{s,T} = k_{E,T}E_{20} \tag{B.10}$$

and

$$f_{y,T} = k_{y,T}f_{y20}, \tag{B.11}$$

respectively. E_{20}, f_{y20} are elastic modulus and yield strength at room temperature, respectively; $k_{E,T}, k_{y,T}$ are reduction factors for elastic modulus and yield strength at elevated temperature, respectively. Table B.1 gives the values for $k_{E,T}, k_{y,T}$. Values for reduction factor for proportional limit are also given in Table B.1.

In ASCE [11], the elastic modulus and yield strength of structural steel at elevated temperature are given by

$$\frac{E_{s,T}}{E_{20}} = 1.0 + \frac{T_s}{2000 In(T_s/1100)} \quad (T_s \leq 600\,^\circ\mathrm{C}) \tag{B.12a}$$

$$\frac{E_{s,T}}{E_{20}} = \frac{690 - 0.69T_s}{T_s - 53.5} \quad (600\,^\circ\mathrm{C} < T_s \leq 1000\,^\circ\mathrm{C}) \tag{B.12b}$$

and

$$\frac{f_{y,T}}{f_{y20}} = 1.0 + \frac{T_s}{900 In(T_s/1750)} \quad (T_s \leq 600\,^\circ\mathrm{C}) \tag{B.13a}$$

$$\frac{f_{y,T}}{f_{y20}} = \frac{340 - 0.34T_s}{T_s - 240} \quad (600\,^\circ\mathrm{C} < T_s \leq 1000\,^\circ\mathrm{C}), \tag{B.13b}$$

respectively.

In CECS200, the elastic modulus and yield strength of structural steel at elevated temperature are given by

$$\frac{E_{s,T}}{E_{20}} = \frac{7T_s - 4780}{6T_s - 4760} \quad (T_s \leq 600\,^\circ\mathrm{C}) \tag{B.14a}$$

$$\frac{E_{s,T}}{E_{20}} = \frac{1000 - T_s}{6T_s - 2800} \quad (600\,^\circ\mathrm{C} < T_s \leq 1000\,^\circ\mathrm{C}) \tag{B.14b}$$

and

$$\frac{f_{y,T}}{f_{y20}} = 1.0 \quad (20\,^\circ\mathrm{C} \leq T_s < 300\,^\circ\mathrm{C}) \tag{B.15a}$$

Table B.1 Reduction factors for structural steel at elevated temperature

Steel temperature	Reduction factor for elastic modulus $k_{E,T} = E_{s,T}/E_{20}$	Reduction factor for yield strength $k_{y,T} = f_{y,T}/f_{y20}$	Reduction factor for proportional limit $k_{p,T} = f_{p,T}/f_{p20}$
20 °C	1.000	1.000	1.000
100 °C	1.000	1.000	1.000
200 °C	0.900	1.000	0.807
300 °C	0.800	1.000	0.613
400 °C	0.700	1.000	0.420
500 °C	0.600	0.780	0.360
600 °C	0.310	0.470	0.180
700 °C	0.130	0.230	0.075
800 °C	0.090	0.110	0.050
900 °C	0.0675	0.060	0.0375
1000 °C	0.0450	0.040	0.0250
1100 °C	0.0225	0.020	0.0125
1200 °C	0.000	0.000	0.000

Note For intermediate values of the steel temperature, linear interpolation may be used

$$\frac{f_{y,T}}{f_{y20}} = 1.24 \times 10^{-8}T_s^3 - 2.096 \times 10^{-5}T_s^2 + 9.228 \times 10^{-3}T_s - 0.2168$$

$$(300\,°C \leq T_s < 800\,°C) \tag{B.15b}$$

$$\frac{f_{y,T}}{f_{y20}} = 0.5 - T_s/2000 \quad (800\,°C \leq T_s < 1000\,°C), \tag{B.15c}$$

respectively.

B.2.2 Stress-Strain Relation

In EC3, the stress-strain relation for structural steel is given by (not considering strain hardening)

$$\sigma_s = \varepsilon E_{s,T} \quad (\varepsilon \leq \varepsilon_{p,T}) \tag{B.16a}$$

$$\sigma_s = f_{p,T} - c + \frac{b}{a}\sqrt{a^2 - (\varepsilon_{y,T} - \varepsilon)^2} \quad (\varepsilon_{p,T} < \varepsilon < \varepsilon_{y,T}) \tag{B.16b}$$

$$\sigma_s = f_{y,T} \quad (\varepsilon_{y,T} \leq \varepsilon \leq \varepsilon_{t,T}) \tag{B.16c}$$

$$\sigma_s = f_{y,T}[1 - \frac{\varepsilon - \varepsilon_{t,T}}{\varepsilon_{u,T} - \varepsilon_{t,T}}] \quad (\varepsilon_{t,T} < \varepsilon < \varepsilon_{u,T}) \tag{B.16d}$$

$$\sigma_s = 0 \quad (\varepsilon = \varepsilon_{u,T}) \tag{B.16e}$$

where

$$a^2 = (\varepsilon_{y,T} - \varepsilon_{p,T})(\varepsilon_{y,T} - \varepsilon_{p,T} + \frac{c}{E_{s,T}}) \qquad \text{(B.17)}$$

$$b^2 = c(\varepsilon_{y,T} - \varepsilon_{p,T})E_{s,T} + c^2 \qquad \text{(B.18)}$$

$$c = \frac{(f_{y,T} - f_{p,T})^2}{(\varepsilon_{y,T} - \varepsilon_{p,T})E_{s,T} - 2(f_{y,T} - f_{p,T})} \qquad \text{(B.19)}$$

here, $\varepsilon_{p,T} = f_{p,T}/E_{s,T}$ $\varepsilon_{y,T} = 0.02$ $\varepsilon_{t,T} = 0.15$ $\varepsilon_{u,T} = 0.20$.
    In ASCE [11], the stress-strain relation for structural steel is given by

$$\sigma_s = \varepsilon E_{s,T} \quad (\varepsilon \leq \varepsilon_{p,T}) \qquad \text{(B.20a)}$$

$$\sigma_s = (12.5\varepsilon + 0.975)f_{y,T} - \frac{12.5f_{y,T}^2}{E_{s,T}} \quad (\varepsilon > \varepsilon_{p,T}) \qquad \text{(B.20b)}$$

where

$$\varepsilon_{p,T} = \frac{0.975f_{y,T} - 12.5f_{y,T}^2}{E_{s,T} - 12.5f_{y,T}} \qquad \text{(B.21)}$$

In CECS200, perfect elastic-plastic relation is used.

Appendix C
Commonds for Modified One Zone Model

```
*creat,PHDcmd,pdan
    /config,nres,10000000
    /config,nproc,2
    /nerr,0,9999999
    /prep7
    !————CONSTANTS————
    stef=5.67e-8 !Stefan-Boltzman constant, J/(sm2K4)
    hfl=35 !film coefficient at fire exposed surface, W/m2K
    hfr=9 !film coefficient at air exposed surface, W/m2K
    Tdefault=9000 !default compute time, s
    !————AIR PROPERTIES————
    rhoa=1.2 !air conductivity, kg/m3
    ca=1000 !air specific heat, J/(kgK)
    ka=0.023 !air conductivity, W/mK
    !————WALL PROPERTIES————
    M=4 !factor controlling numerical accuracy,M=dx*dx/alpha/dt
    rhow=2300 !wall density, kg/m3
    cw=1000 !wall sepcfic heat, J/(kgK)
    kw=1.6 !wall conductivity, W/mK
    emisw=0.8 !wall emissivity
    dw=200/1000 !wall thickness, m
    alphaw=kw/rhow/cw !wall diffusivity
    deltaw=(M*alphaw*10)**(1/2) !mesh size of the wall
    numw=min(8,nint(dw/deltaw))
    !————ROOM GEOMETRIES————
    W=3 !room width, m
    D=4 !room depth, m
    H=2.7 !room height, m
    Vr=W*D*H !room volume, m3
    Af=W*D !floor area, m2
    Wo=1 !open width, m
```

© Springer-Verlag Berlin Heidelberg 2015
C. Zhang, *Reliability of Steel Columns Protected by Intumescent Coatings Subjected to Natural Fires*, Springer Theses, DOI 10.1007/978-3-662-46379-6

```
Ho=2 !open height, m
Ao=Wo*Ho !open area in fire conditions, m2
At=2*(Af+W*H+D*H)-Ao !total area not inclue open, m2
emisr=1-exp(-1.1*H) !gas emissivity
!————HRR————————
alpha=1000/300/300*1000 !fire factor, W/s2
qf=600e6 !fire load per unit floor area, J/m2
qfuelmax=250e3 !fuel controlled max HRRPUA, W/m2
Qmaxfuel=qfuelmax*Af !fuel controlled max HRR, W
Qmaxopen=1500*Ao*Ho**(1/2)*1000 !open controlled max HRR, W
Qmax=Qmaxopen
!Qmax=min(Qmaxfuel,Qmaxopen) !max HRR
t1=(Qmax/alpha)**(1/2) !develop duration, s
Q1=1/3*Qmax*t1 !Heat cnsumed in developng stage, W
t2=(0.7*qf*Af-Q1)/Qmax !steady duration, s
t3=0.6*qf*Af/Qmax !decay duration,s
*dim,HRR,table,3601,1,,time
*do,i,1,3601
HRR(i,0)=(i-1)*10
*if,10*(i-1),le,t1,then
HRR(i,1)=alpha*100*(i-1)*(i-1)
*elseif,10*(i-1),le,t1+t2
HRR(i,1)=Qmax
*else
HRR(i,1)=max(0,Qmax/t3*(t1+t2+t3-10*(i-1)))
*endif
*enddo
!————STEEL PROPERTIES AND GEOMETRIES————
rhos=7850 !steel density, kg/m3
cs=600 !steel specific heat, J/(kgK)
ks=45 !steel conductivity, W/mK
emiss=0.8 !steel emissivity
bs=200/1000 !steel section H200*200*8*12
hs=200/1000
tf=12/1000
tw=8/1000
lc=2.5 !column length, m
Asc=2*bs*tf+(hs-2*tf)*tw !steel cross area, m2
Ass=lc*(2*hs+4*bs-2*tw) !steel surface area,m2
Vs=lc*Asc !steel volume, m3
!————INSULATION PROPERTIES————————
rhoi=300 !insulation density, kg/m3
ci=1200 !insulation specific heat, J/(kgK)
ki=0.12 !insulation conductivity, W/mK
alphai=ki/rhoi/ci !insulation diffusivity
```

```
!di=ki*((30/40/(Tec3free-140))**(1/0.77)*Ass/Vs) !required insulation thickness
for 3h
di=9/1000 !insulation thickness, m
deltai=(M*alphai*10)**(1/2) !mesh size of the insulation
numi=max(4,nint(di/deltai))
Ai=lc*(2*(hs+2*di)+4*(bs+2*di)-2*(2*di+tw))
!————————Tsmax by t-equivalent————————
bound=(rhow*cw*kw)**(1/2)
*if,bound,gt,2500,then
kb=0.04
*elseif,bound,ge,720
kb=0.055
*else
kb=0.07
*endif
teq=qf/1e6*kb*Af/(At+Ao)/(Ao*Ho**(1/2)/(At+Ao))**(1/2) !eqivalent standard
fire exposure, min
Tsteq0=teq/40*((Ass/Vs)/(di/ki))**0.77+140 !Tsmax calculated by using
t-equivalent
Tsteq=-0.0024*Tsteq0*Tsteq0+3.2*Tsteq0-400 !Tsmax calculated by proposed
formula
fini
/prep7
mp,dens,1,rhow !wall
mp,c,1,cw
mp,kxx,1,kw
mp,dens,2,rhoa !air
mp,c,2,ca
mp,kxx,2,ka
mp,dens,3,rhos !steel
mp,c,3,cs
mp,kxx,3,ks
!————————————————————————————————————
mp,dens,4,1e-3 !perfect conductor
mp,c,4,1e-3
mp,kxx,4,1e8
mp,hf,5,hfl !film coefficient at fire exposed surface
mp,hf,6,hfr !film coefficient at air exposed surface
!————————effective covection factor considering heat loss by gas exchange at
opening
mp,hf,7,0.5*Ao*(Ho)**(1/2)*ca/Ao
!————————————————————————————————————
mp,dens,8,rhoi !insulation
mp,c,8,ci
mp,kxx,8,ki
```

```
r,1,At !link32, wall conduction
r,2,At,1,emisr*emisw,stef !link31, fire exposed
r,3,Ao,1,emisr,stef !link31, open radiation
r,4,At !link34, wall convection
r,5,Ao !link34, open convection
r,6,Vr-Vs !mass71,air
r,7,Vs !mass71,steel
r,8,1/0.5/deltaw !link32, HRR
r,9,Ass,1,emisr*emiss,stef !link31
r,10,Ass !link34
r,11,At,1,emisw,stef !link31, wall radiation to air
r,12,Ass !link32
fini /prep7 !thermal analysis
et,1,link32 !conduction bar
et,2,link31 !radiation bar
et,3,link34 !convection bar
et,4,mass71,,,0 !lumped mass
!————MODEL WALL————
n,1,0,0
n,numw+1,dw,0
fill,1,numw+1
type,1
mat,1
real,1
*do,i,1,numw
e,i,i+1
*enddo
!————BC at air exposed———
n,numw+2,dw+deltaw,0
type,2
real,11
e,numw+1,numw+2
type,3
real,4
mat,6
e,numw+1,numw+2
!————BC at fire exposed———
n,numw+3,-deltaw,0
type,2
real,2
e,numw+3,1
type,3
real,4
mat,5
e,numw+3,1
```

```
!————————BC at opening—————-
type,2
real,3
n,numw+4,-deltaw,-deltaw
e,numw+3,numw+4
type,3
mat,7
real,5
e,numw+3,numw+4
!————————HEAT FLUX TO STEEL——
n,numw+5,-deltaw,deltaw
n,numw+5+numi,-deltaw,deltaw+di
fill,numw+5,numw+5+numi
type,1
real,12
mat,8
*do,i,1,numi
e,numw+5+i-1,numw+5+i
*enddo
type,2
real,9
e,numw+3,numw+5
type,3
real,10
mat,5
e,numw+3,numw+5
!————————LUMPED MASS—————-
type,4
mat,2
real,6
e,numw+3
mat,3
real,7
e,numw+5+numi
!————————MODEL HRR—————
n,numw+6+numi,-1.5*deltaw,0
type,1
mat,4
real,8
e,numw+6+numi,numw+3
!————————HEAT LOADING—————
toffset,273
tunif,20
esel,s,mat,,4
bfe,all,hgen,,
```

```
d,numw+2,temp,20
d,numw+4,temp,20
fini
/solu
antype,4
autots,on
deltim,10,1,100
time,Tdefault
outres,nsol,all
allsel
solve
fini
/post26
nsol,2,numw+5+numi,temp
store,merge
*GET,size,VARI,,NSETS
*dim,TsFEM,array,size,1
vget,TsFEM(1),2
*vscfun,TsmaxFEM,max,TsFEM
beta1=Tsteq0-TsmaxFEM
beta2=Tsteq-TsmaxFEM
fini
*end
/inp,PHDcmd,pdan
/pds
pdanl,PHDcmd,pdan
PDVAR,qf,LOG1,600e6,240e6 !PDF of fire load, Gaussian
PDVAR,alpha,GAUS,0.012*1000,0.003*1000 !PDF of growth fire coefficient,
Gaussian
PDVAR,Ao,GAUS,2,2*0.2 !PDF of coefficient considering the opening of win-
dows in fire, TGAU
PDVAR,kw,GAUS,1.6,0.16
PDVAR,dw,GAUS,200/1000,200/1000*0.1
PDVAR,ki,GAUS,0.12,0.12*0.3
PDVAR,di,LOG1,9/1000,9/1000*0.2
PDVAR,Tsteq0,resp
PDVAR,Tsteq,resp
PDVAR,TsmaxFEM,resp
PDVAR,beta1,resp
PDVAR,beta2,resp
PDMETH,MCS,LHS
PDLHS,200,1,,,,,,INIT
PDEXE,MCStri,,1e7
```

Appendix D
Commonds for Monte Carlo Simulations

```
function y=NbT(fy20,e20,T,bs,hs,tw,tf,lc)
    Asc=2*bs*tf+(hs-2*tf)*tw;
    Ixx=(bs*hs^3-(bs-tw)*(hs-2*tf)^3)/12;
    Iyy=(2*tf*bs^3+(hs-2*tf)*tw^3)/12;
    ix=sqrt(Ixx/Asc);
    iy=sqrt(Iyy/Asc);
    lambdax=lc/ix;
    lambday=lc/iy;
    lambda20=lambday;
    lambdaE20=lambda20/3.14*sqrt(fy20/e20);
    if T<100
    ered=1;
    elseif T<500
    ered=-0.001*T+1.1;
    elseif T<600
    ered=-0.0029*T+2.05;
    elseif T<700
    ered=-0.0018*T+1.39;
    elseif T<800
    ered=-0.0004*T+0.41;
    elseif T<1200
    ered=-0.000225*T+0.27;
    else
    ered=0;
    end
    eT=e20*ered;
    if T<=400
    fyred=1;
    elseif T<500
```

© Springer-Verlag Berlin Heidelberg 2015
C. Zhang, *Reliability of Steel Columns Protected by Intumescent Coatings Subjected to Natural Fires*, Springer Theses, DOI 10.1007/978-3-662-46379-6

```
fyred=-0.0022*T+1.88;
elseif T<600
fyred=-0.0031*T+2.33;
elseif T<700
fyred=-0.0024*T+1.91;
elseif T<800
fyred=-0.0012*T+1.07;
elseif T<900
fyred=-0.0005*T+0.51;
elseif T<1200
fyred=-0.0002*T+0.24;
else
fyred=0;
end
fyT=fy20*fyred;
lambdaET=lambdaE20*sqrt(fyred/ered);
betaT=0.65;
alphaT=betaT*sqrt(235/fy20);
phiT=0.5*(1+alphaT*lambdaET+lambdaET^2);
chiT=min(1,1/(phiT+sqrt(phiT^2-lambdaET^2)));
NbT=chiT*Asc*fyT;
y=NbT;
function [A B]=MCS(Wi,Di,Hi,Woi,Hoi,qf0,Ri0,fy20,e20,mu0,Nb0,bs0,hs0,tw0,tf0,lc0)
N=1000000;
qf=lognrnd(log(qf0^2/((qf0*0.3)^2+qf0^2)^0.5),(log((0.3*qf0)^2/qf0^2+1))^0.5,N,1);
Ri=lognrnd(log(Ri0^2/((Ri0*0.3)^2+Ri0^2)^0.5),(log((0.3*Ri0)^2/Ri0^2+1))^0.5,N,1);
zeta=lognrnd(log(0.2^2/((0.2*1)^2+0.2^2)^0.5),(log((1*0.2)^2/0.2^2+1))^0.5,N,1);
b=normrnd(2014,2014*0.09,N,1);
kb=ones(N,1);
for i=1:N
if b(i)>2500
kb(i)=0.04;
elseif b(i)<=2500
kb(i)=0.055;
else
kb(i)=0.07;
end
end
W=Wi.*ones(N,1);D=Di.*ones(N,1);H=Hi.*ones(N,1);
Wo=Woi.*ones(N,1);Ho=Hoi.*ones(N,1);
Af=W.*D;At=2.*(W.*D+W.*H+D.*H);Ao=Wo.*Ho;
OF=Ao.*Ho.^0.5./At;
```

```
wf=(OF.*(1-zeta)).^(-0.5).*Af./At;
teq=qf.*kb.*wf;
lc=normrnd(lc0,lc0*0.05,N,1);
fy20=normrnd(fy20,fy20*0.063,N,1);e20=normrnd(e20,e20*0.045,N,1);
bs=normrnd(bs0,bs0*0.05,N,1);
hs=normrnd(hs0,hs0*0.05,N,1);
tw=normrnd(tw0,tw0*0.05,N,1);tf=normrnd(tf0,tf0*0.05,N,1);
C=4.*bs-2.*tw+2.*hs;As=2.*bs.*tw+tw.*(hs-2.*tf);AiV=C./As.*1000;
Iyy=(2.*tf.*bs.^3+(hs-2.*tf).*tw.^3)./12;
iyy=(Iyy./As).^0.5;
lambday=lc./iyy;
lambdaE=lambday./3.14.*(fy20./e20).^0.5;
alpha=0.65.*(235./fy20).^0.5;
phi=(1+alpha.*lambdaE+lambdaE.^2)./2;
chi=1./(phi+(phi.^2-lambdaE.^2).^0.5);
Nb=chi.*As.*fy20;
PT=normrnd(mu0*Nb0,mu0*Nb0*0.3,N,1);
mu=PT./Nb;
PF1=lognrnd(-0.053455,0.121189,N,1);
PF2=evrnd(1.00378,0.0954518,N,1);
Tb=PF2.*(39.19.*log(1/0.9674./mu.^3.833-1)+482);
Delta=teq./40.*(AiV./Ri).^0.77;
Tsmax=PF1.*(-0.0024.*Delta.^2+2.528.*Delta+0.96);
n=0;m=0;
for i=1:N
if Tsmax(i)>=300&&Tsmax(i)<=600
if Tb(i)<=Tsmax(i)
n=n+1;
else
n=n;
end
m=m;
else
m=m+1;
end
end
Pfail=n/(N-m);
Beta=norminv(1-Pfail,0,1);
A=Pfail;
B=Beta;
```

References

1. J. Holman, *Heat Tranfer*, 6th edn. (McGraw-Hill, New York, 1986)
2. D. Drysdale, *An Introduction to Fire Dynamics*, 2nd edn. (Wiley, New York, 1999)
3. J. Rockett, J. Mike, SFPE Handbbok of Fire Protection Engineering Society of Fire Protection Engineers, 2002), 3rd edn., chap. Section 1–2: Condction of Heat in Solids
4. A. Jowsey, *Fire Imposed Heat Fluxes For Structural Analysis*. Ph.D. thesis (The University of Edinburgh, 2006)
5. B. McCaffrey, J. Quintiere, M. Harkleroad, Fire Technology **17**(2), 98 (1981)
6. J.H. Lienhard IV, J.H. Lienhard V, *A Heat Transfer Textbook*, 3rd edn. (Phlogiston Press, Cambridge, Massachusetts, 2003)
7. R. Siegel, J. Howell, Thermal Radiation Heat Transfer, 3rd edn. (Hemisphere Publishing Corporation, Washington DC, 1992)
8. BSI, *Eurocode 4: Design of Composite Steel and Concrete Structures—Part 1–2: General Rules—Structural Fire Design* (British Standard, 2005)
9. M.F. Modest, *Heat Transfer Handbook* (Wiley, New York, 2003), (chap. Section 8: Thermal radiation)
10. BSI, *Eurocode 3: Design of Steel Structures—Part 1–2: General Rules—StrUctural Fire Design* (British Standard, 2005)
11. ASCE, *Structural Fire Protection* (ASCE Committee on fire protection, Manual No.78, Reston, VA, 1992)
12. China Association for Engineering Construction Standardization (CECS200), *Technical Code for Fire Safety of Steel Structure in Buildings* (Beijing, China Planning Press, 2006)

Printed in the United States
By Bookmasters